"An examination of the remarkable role of the shadowy but powerful 'amateur scientist' whose intellect and energy spurred critical scientific research that shortened and helped win WWII. . . . Remarkable and remarkably told, as if F. Scott Fitzgerald had penned *Batman*."

—*Kirkus Reviews*

"By the time you are finished, you are prepared to bestow on Alfred Lee Loomis the title of Most Interesting Man I Never Knew Anything About. . . . Loomis and Conant are just right for each other."

—Alex Beam, *The New York Times Book Review*

"[Conant's] group portrait offers a healthy reminder of how much good science depends on community and collaboration, not solitary genius."

—*The New Yorker*

"An eccentric, fabulously wealthy scientist performs groundbreaking experiments on the nature of time in his stone castle and, after hosting a sumptuous feast for his colleagues and friends, forces his guests to participate in brain-wave experiments while hypnotized. Something out of H. G. Wells or Mary Shelley? No, a real scene from the life of Alfred Lee Loomis, the extraordinary American financier, scientist, and philanthropist who played a pivotal role in the development of radar and the creation of the Manhattan Project during World War II. Jennet Conant . . . has written a fascinating biography of this unusual and impressive figure."

—Richard Di Dio, *The Philadelphia Inquirer*

"A must-read for fans of World War II history, and it will captivate students of science and technology."

—Otis Port, *BusinessWeek*

"More than a vivid biography of Alfred Lee Loomis, this is a bright and intelligent portrait of a season of science in America that changed history."

—*Library Journal*

"Like the character of Loomis himself, this is a fabric woven of many strands—financial genius, brilliant inventiveness, a passion for science, human traits and appetites—each essential to the emergent pattern. . . . Thanks to Conant's efforts, the tapestry is at last on display."

—Fred Bortz, *Pittsburgh Post-Gazette*

"This is a very good book. . . . Once you start it, you will have a hard time putting it down."

—Jeremy Bernstein, *The Washington Times*

"In *Tuxedo Park*, Conant has indeed written a fascinating tale."

—Jules Wagman, *Milwaukee Journal Sentinel*

"Alfred Loomis has remained deep in the shadows of history until now. . . . Riveting."

—Joseph Losos, *St. Louis Post-Dispatch*

"It's a tale that sounds more like an Ian Fleming creation than truth: An eccentric tycoon brings the world's brightest minds to a private enclave, where they develop inventions that alter world history. . . . Conant praises the financier-turned-scientist's work and . . . also captures a tarnished image of Loomis the man. . . . Conant resurrects the explosive contributions of a man who harbored little interest in being remembered."

—Stuart Wade, *Austin American Statesman*

"No one man won World War II for us, but none exceeded Alfred Loomis's contribution. He was critical to crucial developments, everything from radar to the atomic bomb. He put into victory his genius, his energy, and his Wall Street fortune. Author Jennet Conant put all of her considerable talents into this biography, which is as superb as the subject."

—Stephen E. Ambrose

"Alfred Lee Loomis, who lived among the swells in a gated Tuxedo Park, hated F.D.R., rarely communicated with his wife and three sons, stole his best friend's wife, and with icy disdain helped drive an aide to take his own life. Yet the Allies may not have won World War II without this man whom history forgot. As Jennet Conant's heart-thumping book recounts, Loomis was a public-spirited citizen with the brilliance and ability to galvanize the scientific community to invent first the potent weapon that came to be called radar to spare London from bombs and to destroy U-2 boats, and later contributed to the making of the atom bomb. Long after you race to the end, this heroic story will linger in memory."

—Ken Auletta

Jennet Conant

TUXEDO

PARK

*A Wall Street Tycoon and the Secret
Palace of Science That Changed
the Course of World War II*

SIMON & SCHUSTER

New York   London   Toronto   Sydney   Singapore

SIMON & SCHUSTER
Rockefeller Center
1230 Avenue of the Americas
New York, NY 10020

First Simon & Schuster trade paperback edition 2003

SIMON & SCHUSTER and colophon are registered trademarks
of Simon & Schuster, Inc.

For information regarding special discounts for bulk purchases,
please contact Simon & Schuster Special Sales at
1-800-456-6798 or business@simonandschuster.com

Designed by Karolina Harris

Manufactured in the United States of America
1   2   3   4   5   6   7   8   9   10

The Library of Congress has cataloged the hardcover edition as follows:
Conant, Jennet.
Tuxedo Park : a Wall Street tycoon and the secret palace of science
that changed the course of World War II / Jennet Conant.
p.   cm.
Includes bibliographical references and index.
1. Loomis, Alfred L. (Alfred Lee), 1887–1975. 2. Physicists—United States—Biography.
3. Atomic bomb—United States—History—20th century.
4. Research—New York (State)—Tuxedo Park—History—20th century.
5. World War, 1939–1945—Science—United States.   I. Title.
QC16.L647 C66   2002
530'.0092'274731—dc21   2002021001   [B]

ISBN 0-684-87287-0
0-684-87288-9 (Pbk)

For Steve and John

To advance scientific knowledge, pick a man of genius, give him money, and let him alone.

—James B. Conant

You don't know what life really is
Till you've been to Tuxedo Park!

—Chorus of a song from the musical *Tuxedo*
by Henry J. Sayers, 1891

Contents

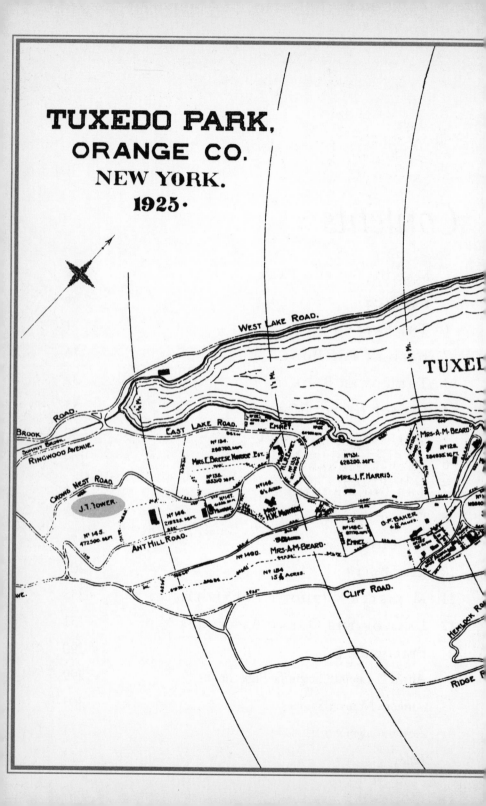

Preface

I F the phrase *stranger than fiction* ever described any series of events, it applies to the bizarre circumstances that surrounded the suicide of William Richards on January 30, 1940, on the eve of the publication of his novel, *Brain Waves and Death*. The book, which was written under a pseudonym, was a thinly veiled account of the legendary scientific laboratory owned by the millionaire Alfred Lee Loomis and the eccentric coterie of geniuses whose work he financed. Richards was an accomplished chemist and had for years enjoyed Loomis' luxurious facilities in the exclusive enclave of Tuxedo Park, where his mansion was known to be a meeting place for the great names in science and finance. Richards, who was my great-uncle, came from a prominent Boston family and was painfully aware of his pedigreed seat in the country's intellectual elite. His father, Theodore William Richards, was chairman of the Harvard Chemistry Department and a Nobel laureate. His sister, Grace Richards, was married to James B. Conant, who at the time was president of Harvard. In this rarefied company, it was not enough to be merely accomplished—anything less than extraordinary constituted a disappointment. Richards' talents lay in music and art, but he was expected to strive for greatness in science. Before he turned forty, deciding he had fallen short of the mark, he killed himself. Within the Richards-Conant family, his suicide was regarded as a kind of weakness, a moral failure. It was not only a betrayal of his intellectual promise, but an embarrassing public expression of his private anguish. My grandfather used his influence to have the incident covered up, and it was never spoken of again.

My father, Theodore Richards Conant, knew only that he had lost his favorite uncle to some terrible tragedy. The deep air of mystery that surrounded Richards' death, and his fiction's rich and foreboding scientific detail, always haunted my father. He saved a copy of the scandalous novel, which was published posthumously and quickly disappeared, along with an unpublished short story about a scientist working to create the first atomic bomb, which my grandfather confiscated on the grounds that it was too close to the truth to dare publish in those dangerous times. The silence my grandfather imposed served only to distort and enlarge the family myths about William Richards, and in my father's boyish eyes, he became a heroic figure—rebellious, romantic, doomed.

Years later, when I was growing up, my father liked to tell lurid tales about Richards' death. At funerals, usually held at the family plot in Mount Auburn Cemetery in Cambridge, Massachusetts, we would go hunt for his grave and that of his brother, who also committed suicide. My father always joked that every good Boston family should have a pew at St. Trinity's, a plot at Mount Auburn's, and a gurney at McLean's (the local nuthouse). Once, when we were visiting 17 Quincy Street, the Harvard president's house where he was raised, my father took me to an upstairs bedroom, pointed to the four-poster, and told me that this was where his uncle's body had been found. He fed my fascination by telling me vivid, and wildly varying, accounts of what had happened. In one version, Richards had built an elaborate apparatus that he used to electrocute himself, and it was my horrified grandmother who discovered the corpse. Another time, he told my grandfather's Harvard biographer that the contraption had been "rigged to an alarm clock that released a lethal dose of poison gas," killing Richards exactly the same way the characters in his novel were finished off. Neither account was accurate, but the biographer believed him, and so did I.

Over the years, I became enchanted with the stories and the extraordinary coincidence of his suicide, the approaching war, and the invention of the bomb. That my grandfather played a crucial role in the decision to build the first atomic bomb, and administered the Manhattan Project, only added to the mystery. How had Richards come to know so much about something as secret as nuclear fission? Did he know too much? Had he exposed more than he had imagined in his roman à clef about Loomis' private laboratory? As I grew older, I pried open locked trunks stored in the basement of my grandparents' country home in Hanover, New Hampshire, and pored over old letters and diaries, looking for clues. When my father gave a large cache of books and papers belonging to my grandfather to Harvard in the mid-1980s, I asked him not to hand over any of Richards' letters. I was an adult, on the brink of a journalism career, and I knew with certainty that one day I would write about this strange chapter in my family's history.

At the same time, I knew it meant acknowledging the strain of manic depression that has been passed down through successive generations of the Richards and Conant families and has been the cause of so much tragedy and pain. I would also be acknowledging my own genetic vulnerability, and it was many years before I felt ready to do that. I also struggled with the problem of prying into what many of my grandfather's friends and colleagues might regard as a dark corner of his illustrious career. James Conant was a very private, proud, and tidy man and placed a premium on appearances. He would have loathed seeing his family's mess tipped onto the page. There were also gaping holes in the story. My grandmother was acutely aware that graduate students would one day paw her private papers, and she set about methodically destroying anything incriminatingly personal in the record, ripping pages out of diaries and burning most of her mother's and brother's letters.

It was ironic, then, that I first looked into Alfred Loomis to help shed light on William Richards' life. Few men of Loomis' prominence and achievement have gone to greater lengths to foil history. Most of the books I looked in for information about Loomis included only a few lines about him, at most a paragraph or two. He seemed to stand at the edge of important events, intimately involved and at the same time somehow overlooked. Yet here was a character who was at once familiar. Independently wealthy, iconoclastic, and aloof, Loomis did not conform to the conventional measure of a great scientist. He was too complex to categorize—financier, philanthropist, society figure, physicist, inventor, amateur, dilettante—a contradiction in terms. At a time when the world was a smaller place, and the men in positions of power all knew one another because of family connections, school ties, and club affiliations, Loomis knew everyone. He was the ultimate insider. Although he rose to become one of the most powerful figures in banking in the 1920s, and scooped his peers by pulling out of the market before the Crash of 1929 and rode out the Depression sitting on a mountain of cash, he was not satisfied with the lucre and laurels of Wall Street. He, too, by virtue of his background and education, felt obliged to strive for a kind of excellence that had nothing to do with the external trappings of success.

Loomis had the foresight to know that science would soon become a dominating force, and he used his immense fortune to attract a gifted group of young physicists to his private laboratory and endow pioneering research that pushed at the frontiers of knowledge. He created a scientific idyll in the cloistered fiefdom of Tuxedo Park, and in his belief in invention and experimentalism, he prepared the way for a series of scientific developments that would not only change the course of the war, but ultimately transform the modern world. For more than a decade, William Richards was part of Loomis' brilliant circle at Tuxedo Park, and in his fiction, he captured that twilight period between the wars when the last of the gentleman scientists engaged in pure research, before the demands of the real world called them to action. In the intersection of their lives, I glimpsed a story of real interest, authentically American, with the stature of history. And in the recurring mental illness that ravaged Loomis' family, and resulted in a bitter divorce that drove him into seclusion, I recognized a parallel story that helped me make peace with my own peculiar legacy.

TUXEDO PARK

Chapter 1

THE PATRON

Ward was smiling but that did not mean that he was amused.
The smile was a velvet glove covering his iron determination
to get under way without any lost motion.

—WR, from *Brain Waves and Death*

O N January 30, 1940, shortly after ten P.M., the superintendent of
the building at 116 East 83rd Street noticed that a bottle of milk delivered that morning to one of his tenants had remained in front of the
door all day. The young man who rented the three-room apartment had
not said anything about going out of town. He was a conspicuous fellow,
extremely tall—at least six feet four—and lean, with piercing blue eyes
and a shock of dark hair. After knocking repeatedly and failing to get an
answer, the superintendent notified the police.

William T. Richards was found dead in the bathtub with his wrists
slashed, blood from his wounds garlanding the walls of the bathroom.
He was dressed in his pajamas, his head resting on a pillow. A razor blade
lay by his hand. He was a former chemistry professor at Princeton University who was currently employed as a consultant at the Loomis Laboratory in Tuxedo Park, New York. He was thirty-nine years old. His
personal papers mentioned a mother, Miriam Stuart Richards, living in
Massachusetts, and the detective at the scene asked the Cambridge po-

lice to contact her. As *The New York Times* reported the following morning, William Richards was from a prominent Boston family, son of the late professor Theodore William Richards of Harvard, winner of a Nobel Prize in chemistry, and the brother of the former Grace (Patty) Thayer Richards, wife of the president of Harvard, James B. Conant.

Although his death was clearly a suicide, everything possible was done to hush up the more unpleasant aspects of the event, and the Boston papers never published the details. Richards' brother, Thayer, was immediately dispatched to New York, and he saw to it that most of what had transpired was concealed from his mother and sister. A suicide note that was found by the tub was destroyed, and its contents were never revealed. The Richards family was naturally concerned about its reputation, but there were also pressing concerns, of a rather delicate nature, that made it vitally important that Bill's suicide be kept as quiet as possible. Miriam Richards, desperate to avoid any scandal, drafted a reassuring letter attempting to put the untimely death of her son in a better light, copies of which she sent out to important friends and relations. She explained that Bill had long been "nervously, seriously ill" and had never properly recovered from severe abdominal surgery several years earlier. She also supplied him with an end that left open the possibility that his death was accidental, writing that "Bill died of an overdose of a sleeping draught." It is entirely possible that this is what she had been told.

"William Theodore Richards was beyond any doubt one of the most brilliant members of our class," began his Harvard obituary, based on the fond reminiscences of his friends and scientific colleagues. He was interested in new scientific phenomena, the originality of his ideas leading him into experimental work. But he had the kind of restless, wide-ranging intelligence—he was a talented painter and musician and briefly considered playing the cello professionally—that made him, according to one friend, "a veritable Renaissance man." He was a chemist at his father's insistence, but his heart was not in it, and he found it difficult to force himself to undertake the routine proofs and laborious accumulation of data that would have given him more publishable material and more recognition in his field. He had "a mentality which could be called great," wrote his classmate Leopold Mannes, a fellow scientist and musician, who speculated that Richards despaired of ever meeting the onerous demands he imposed on himself. "In his attitude towards life, towards science, towards music—of which he had an astounding knowledge and perception—and towards literature, he was a relentless perfectionist, and thus his own implacable judge. No human being could be expected fully to satisfy such standards."

Richards was a solitary man, confining his friends to a small, clever circle. He kept most of his contemporaries at bay with his caustic wit, which made quick work of any human frailty, whether at his own expense or someone else's. With complete abandon, he would ruthlessly mimic anyone from Adolf Hitler to some sentimental woman who had been foolish enough to confide in him. To most, he seemed cordial, cold, and a bit superior, his moodiness exacerbated by periods of poor health and depression. He eventually quit his job at Princeton and moved to New York, where he worked part-time as a chemical consultant while devoting himself to an arduous course of psychotherapy. The Harvard memorial notes concluded that "after a brave struggle for ten years to overcome a serious neurosis, which in spite of treatment grew worse, Bill died by his own hand."

Richards' death was nevertheless "shocking" to Jim Conant and his wife, Patty. Richards had celebrated Christmas with them only a few weeks before and had stayed in the large brick mansion at 17 Quincy Street that was the official residence of the Harvard president. Although his psychological condition had always been precarious, he had seemed "to be making real progress," his mother later lamented in a letter to a close family friend, so much so that "last summer and autumn he was so happy and well that for fun he wrote a detective story." Richards had submitted the manuscript to Scribner's, which "had at once accepted it."

Just a few weeks after he took his own life, his book, *Brain Waves and Death*, was published under the pseudonym "Willard Rich." It was, in most respects, a conventional murder mystery, with the added interest of being set in a sophisticated modern laboratory, where a group of eminent scientists are hard at work on an experiment designed to measure the electrical impulses sent out by the brain. In a twist on the standard "hermetically sealed room" problem, Richards staged the murder in a locked experimental chamber that is constantly monitored by highly sensitive listening devices and a camera. The book earned respectful reviews, with *The New York Times* describing the story as "ingeniously contrived and executed" and awarding Willard Rich "an honorable place in the ranks of mystery mongers." None of the critics were apparently aware that the author was already dead or that he had rather morbidly foreshadowed his imminent demise in the book, in which the first victim is a tall, arrogant young chemist named Bill Roberts.

At the time, only a small group of elite scientists could have known that while the method Richards devised to kill off his literary alter ego was of his own invention—a lethal packet of poison gas that was frozen solid and released into the atmosphere when warmed to room temperature—the actual science and the laboratory itself were real. George Kistiakowsky, a Harvard chemistry professor and one of Richards' closest friends and professional colleagues, guessed the truth immediately, "that it was a take-off on the Loomis Laboratory and the characters frequenting it." Despite its contrived plot, the book was essentially a roman à clef. No one who had ever been there could fail to recognize that the "Howard M. Ward Laboratory" was in reality the Loomis Laboratory in Tuxedo Park and that the charismatic figure of Ward himself was transparently based on Alfred Lee Loomis, the immensely wealthy Wall Street tycoon and amateur physicist who, among his myriad inventions, claimed a patent for the electroencephalograph, a device that measured brain waves.

The opening paragraphs of the book perfectly captured Loomis' rarefied world, where scientists mingled with polite society and where intellectual problems in astronomy, biology, psychiatry, or physics could be discussed and pursued in a genteel and collegial atmosphere:

> The Howard M. Ward Laboratory was not one of those hospital-like institutions where Pure Science is hounded grimly and humorlessly as if it were a venomous reptile; the grounds of the Laboratory included a tennis court, bridle paths, and a nine-hole golf course. Guests there did not have to confine themselves to science, they could live fully and graciously.

It was Richards who had first told Kistiakowsky about Loomis' private scientific playground in Tuxedo Park, a guarded enclave of money and privilege nestled in the foothills of the Ramapo Mountains. Tuxedo Park, forty miles northwest of New York City, had originally been developed in 1886 by Pierre Lorillard, the tobacco magnate, as a private lakefront resort where his wealthy friends could summer every year. The rustic retreat became the prime meeting ground of American society, what Ward McCallister famously called "the Four Hundred," where wealthy moguls communed with nature in forty-room "cottages" with the required ten bedrooms, gardens, stables, and housing for the small army of servants required for entertaining in style. Leading members of the financial elite, such as Astors and Morgans, numbered among the Tuxedo Club's first members, as did Averell Harriman, who occupied a vast neighboring estate known as Arden. Over the years, Tuxedo Park, with its exclusive clubhouse and fabled balls, had taken on all of the luster and lore of a royal court, and although it had dimmed somewhat since the First World War, it still regarded itself as the Versailles of the New York rich.

Loomis, a prominent banker and socialite, was very much part of that world and owned several homes there. According to Richards, however, Loomis was also somewhat eccentric and disdained the glamorous swirl around him. He had developed a passion for science and for some time had been leading a sort of double life: as a partner in Bonbright & Co., the thriving bond investments subsidiary of J. P. Morgan, he had amassed a substantial fortune, which allowed him to act as a patron somewhat in the manner of the great nineteenth-century British scientists such as Charles Darwin and Lord Rayleigh. To that end, Loomis had purchased an enormous stone mansion in Tuxedo, known as the Tower House, and turned it into a private laboratory where he could give free rein to his avocation—primarily physics, but also chemistry, astronomy, and other ventures. He entertained lavishly at Tower House and invited eminent scientists to spend long weekends and holidays as his guests. More to the point, as Richards told Kistiakowsky, Loomis also extended his hospitality to "impecunious" young scientists, offering them stipends so they could enjoy elegant living conditions while laboring as skilled researchers in his laboratory.

Richards had seen to it that Kistiakowsky—"Kisty" to his pals—secured a generous grant from the Loomis Laboratory. The two had met and become fast friends at Princeton in the fall of 1926, when as new chemistry teachers they were assigned to share the same ground-floor laboratory. They were both tall, physically imposing men, with the same contradictory mixture of witty raconteur and reserved, introspective scientist. In no time they had discovered a mutual fondness for late night philosophizing and bathtub gin. As this was during Prohibition, the Chemistry Department had to sponsor its own drinking parties, and the two chemists "doctored" their own mixture of bootleg alcohol and ginger ale with varying degrees of success. Richards, who was subsidized by his well-heeled Brahmin family, had soon noticed that his Russian colleague, a recent émigré who sent money to his family in Europe, was having difficulty managing on the standard instructor's salary of $160 a month. Knowing any extra source of funds would be welcome, Richards had put in a good word with Loomis, just as he had when recommending Kistiakowsky to his "uncle Lawrence"—A. Lawrence Lowell, who was then president of Harvard, and a close family friend. Grinning into the phone, he had provided assurances that Kistiakowsky was not some "wild and woolly Russian" and, despite being just off the boat, was "wholly a gentleman, had proper appearance and table manners, etc."

Richards' own introduction to Loomis had happened quite by accident a few months prior to his arrival at Princeton. While Richards was completing his postdoctoral studies at Göttingen, he had been sitting in the park one Sunday morning, idly reading *Chemical Abstracts*, when a paragraph briefly describing an experiment being carried on in the "Loomis Laboratory" had caught his eye. He had immediately sent off a letter to the laboratory, "suggesting that certain aspects of the experiment could be further developed," and he had even outlined what the result of this development would probably be. Some months later, he received a response from the laboratory informing him that they had carried out his suggestions and the results were those he had anticipated. This had been followed by a formal invitation to work at the Loomis Laboratory.

Over the years, Richards and Kistiakowky had often commuted from Princeton to Tuxedo Park together on weekends and holidays and had conducted some of their research experiments jointly. Richards had arranged for them both to spend the summer of 1930 as research fellows at the Loomis Laboratory. What a grand time that had been. Not only was the room and board better than that of any resort hotel, but weekend recreation at Tower House—when the restriction against women was relaxed—included festive picnics, drinks, parties, and elaborate black-tie dinners. Back then, they had both been ambitious young chemists at the beginning of their careers and had reveled in the chance to work with such legendary figures as R. W. Wood, the brilliant American experimental physicist from Johns Hopkins, whom Loomis had lured to Tuxedo Park as director of his laboratory. Working alongside Loomis and a long list of distinguished collaborators, they had carried out series of original experiments, including some of the first with intense ultrasonic radiation, and had proudly seen their lines of investigation published in scientific journals and taken up by laboratories in America and Europe.

Kistiakowsky, who by then had joined Harvard's Chemistry Department and become close friends with Conant, never publicly revealed that Richards' book was based on Loomis and the brain wave experiments conducted at Tower House. In his carefully composed entry in Richards' Harvard obituary, he made only a passing reference to a "Mr. A. L. Loomis of Tuxedo Park," diplomatically noting that Richards' work at the laboratory had afforded him "one of the keenest scientific pleasures of his career." However, it is typical that he could not resist dropping one hint. Observing that very few physical chemists possessed his late friend's keenness of mind, Kistiakowsky concluded that no one could ever match Richards' own concise presentation of his work, "which was always done in the best literary form."

At the time of Richards' death, Kistiakowsky was still working for Loomis on the side. But the stakes were much higher now, and the project he had undertaken was so secret, and of such fearful importance, that Richards' parody of the Loomis Laboratory must have struck him as a wildly precipitous and ill-conceived prank. Richards had always thumbed his nose at authority and convention and had been disdainful of the narrow scope of his scientific colleagues, whom he once complained talked about "nothing but the facts, the fundamental tone of life, while I prefer the inferred third harmonic." But for Kistiakowsky, a White Russian who at age seventeen had battled the advancing Germans at the tail end of World War I, and then fought the Bolsheviks before being wounded and forced to flee his country, the prospect of another European war took precedence over everything. While in the past he might have joined Richards in poking fun at Loomis and his collector's attitude toward scientists, Kistiakowsky now appreciated him as a man who knew how to get things done. Loomis was a bit stiff, with the bearing of a four-star general in civilian clothes, but he was strong and decisive.

Kistiakowsky did not have to be told to be discreet, though he may have been. Loomis was furious about the book and threatened to sue for libel. He was an intensely private man and was horrified at the breach of trust from such an old friend. Richards had been a regular at the Tower House for more than ten years and was intimately acquainted with the goings-on there. In the months directly preceding his suicide, Loomis had plunged the laboratory into highly sensitive war-related research projects. Loomis wanted no part of the gossip and notoriety that might result either from Richards' unfortunate death or his book.

Neither did Jim Conant, who regarded the book as a source of acute embarrassment. It was bad enough that his wife's family continuously vexed him with their financial excesses and emotional crises, here was his brother-in-law stirring up trouble from the grave with this incriminating tale. Patty Conant was so distressed that she begged her brother, Thayer, to have the book recalled at once. But it was too late for that, and it was not long before Conant discovered that *Brain Waves and Death* was not Richards' only legacy.

With his instinctive ability to home in on the latest developments on the frontiers of research, Richards had followed up his first book with something far more sensational. Among the papers collected from his apartment after his death was the draft of a short story entitled "The Uranium Bomb." It was written once again under the pseudonym Willard Rich. The slim typed manuscript, bearing the name and address of his literary agent, Madeleine Boyd, on the front cover, was clearly intended for publication. Richards was an avid reader of *Astounding Science Fiction* and probably intended to place his story in the magazine, which regularly carried the futuristic visions of H. G. Wells and was a popular venue for the doomsday fantasies of scientists who were themselves good writers. Richards' story opens with the meeting in March 1939 between a rather callow young chemist named Perkins (Richards) and a Russian physicist named Boris Zmenov, who tries to enlist the well-connected American to warn his influential friends, and ultimately the president, "to suppress a threat to humanity." The Zmenov character, who is convinced the Nazis want to build a bomb, explains that there had been a breakthrough in atomic fission: the uranium nucleus had been split up, with the liberation of fifty million times as much energy as could be obtained from any other explosive. "A ton of uranium would make a bomb which could blow the end off Manhattan island."

Richards outlined Zmenov's theory, "tossed off with the breezy impudence of a theoretical physicist," describing the principles of atomic fission and the chain reaction by which an explosion spreads from a few atoms to a large mass of material, thereby generating a colossal amount of power. When Perkins professes disbelief, Zmenov becomes furious: "I am on the verge of developing a weapon," he declares, "which will be the greatest military discovery of all time. It will revolutionize war, and make the nation possessing it supreme. I wish that the United States should be this nation, but am I encouraged? Am I assisted with the most meager financial support? Bah."

As Conant read the manuscript, he realized it was an accurate representation of the facts as far as they were known. While not exactly common knowledge, Conant was aware that a great deal of information about uranium had been leaking out in scientific conferences and journals over the past year. His brother-in-law could have easily picked up many of his ideas just from reading *The New York Times*, which had extensively covered the lecture appearances of the Danish physicist Niels Bohr and his outspoken remarks about the destructive potential for fission. Even *Newsweek* had reported that atomic energy might create "an explosion that would make the forces of TNT or high-power bombs seem like firecrackers." For his part, Conant, an accomplished scientist who had been chairman of Harvard's Chemistry Department before becoming president of the university, was far from convinced atomic fission was anywhere near to being used as a military weapon. He was still inclined to believe the only imminent danger from fission was to some university laboratories. But he was not ready to dismiss it, either.

Richards' story was disturbing, and if it cut as close to the bone as his novel had, it was potentially dangerous. There were too many familiar names for comfort, including an acquaintance "prominent in education circles" by the name of "Jim," which Conant must have read as a sly reference to himself. More troubling still, the physical description of Zmenov—very short, round, and excitable—matched that of the Hungarian refugee scientist Leo Szilard, who was known to be experimenting with uranium fission at Columbia University in New York. Szilard was always agitating within the scientific community about the importance of fission and had even formed his own association to solicit funds for his work. In a scene that rang especially true, Perkins arranges for Zmenov to meet a wealthy banker, and Zmenov is crestfallen when he does not pull out his checkbook. "Perhaps Zmenov thought all bankers were crazy to find something to sling their money into," Richards wrote in yet another thinly disguised account of Loomis' exploits. This time, Harvard's cautious president did not wait for Loomis to tell him that the story revealed too great a knowledge of high-level developments in the scientific world, and at the very moment external pressures were coming to a peak. Conant made sure the story was suppressed.

Conant was too guarded to ever fully confide his doubts in anyone, but he expressed some of his reservations to his son, Ted, who was thirteen years old at the time. The boy had come across the story when going through the boxes of books and radio equipment Richards had left to him and insisted that it ought to be published according to the wishes of his beloved uncle. Anything short of that, he argued, "was censorship." The fierce row between father and son that followed was memorable because it was so rare. Conant was a calm, controlled man who rarely lost his temper. He was also coldly practical and not given to old-fashioned sentiment. His angry retort that Richards' story was "outlandish" and "unworthy of him," coupled with his uncharacteristic claim that "the family honor was at stake," suggested there was something more to his opposition than he was letting on. His son reluctantly let the matter drop.

By the time Conant discovered Richards' manuscript, many of the events described in the story, although slightly distorted, had in fact already transpired. Szilard had befriended Richards and was regularly updating him on the work he was carrying on with the Italian émigré physicist Enrico Fermi, who had won a Nobel Prize and had recently joined the staff of Columbia University. After the French physicist Frédéric Joliot-Curie published his findings on uranium fission, Fermi lost patience with Szilard's passion for secrecy and insisted that their recent experiments be published. In a hasty note to Richards on April 18, 1939, Szilard broke the news:

> Dear Richards:—
>   It has now been decided to let the papers come out in the next
> issue of *Physical Review*, and I wanted you to be informed of this fact.
>   With kind regards,
>   yours,
>   [Leo Szilard]

As Richards cynically noted in his story, Szilard's interest in him was primarily as a link to private investors like Loomis, whom Szilard desperately wanted to bankroll the costly experiments he planned to do at Columbia University. At the same time, Szilard had been busy wooing other Wall Street investors, enticing them with the promise of cheap energy. In a letter to Lewis L. Strauss, a New York businessman interested in the atom's commercial potential, Szilard wrote tantalizingly of "a very sensational new development in nuclear physics" and predicted that fission "might make it possible to produce power by means of nuclear energy." At one point, Szilard arranged for himself and Fermi to have drinks at Strauss' apartment and asked Strauss to invite his wealthy acquaintance Lord Rothschild, but the two physicists could not persuade the English financier to underwrite their chain reaction research. Part of the problem was that while Szilard needed backers, he was desperately afraid Germany would realize fission's military potential first. He was obsessed with secrecy. He was determined to protect his discoveries and cloaked his project in so much mystery that he often appeared as "paranoid" as Richards portrayed him in his sharp caricature. After all his efforts to find private investors had met with failure, Szilard wrote to Richards on July 9, 1939, pleading for money to prove "once and for all if a chain reaction can be made to work." His tone was urgent:

Dear Richards:

I tried to reach you at your home over the telephone, but you seemed to be away, and so I am sending this letter in the hope that it might be forwarded to you. You can best see the present state of affairs concerning our problem from a letter which I wrote to Mr. Strauss on July 3rd, a copy of which I am enclosing for your information and the information of your friends. Not until three days ago did I reach the conclusion that a large scale experiment ought to be started immediately and would have a good chance of success if we used about $35,000 worth of material, about half this sum representing uranium and the rest other ingredients. . . . I am rather anxious to push this experiment as fast as possible. . . . I would, of course, like to know whether there is a chance of getting outside funds if this is necessary to speed up the experiment, and if you have any opinion on the subject, please let me know.

If you think a discussion of the matter would be of interest I shall of course be very pleased to take part in it. . . . Please let me know in any case where I can get hold of you over the telephone and your postal address.

During the summer of 1939, Szilard and Fermi worked out the basis for the first successful chain reaction in a series of letters. Encouraged by their correspondence, but frustrated by his continued failure to enlist any financial support for his experiments, Szilard turned to his old mentor, Albert Einstein, for help. Einstein was sixty years old and famous, someone with enough stature to lend credibility to his cause. After meeting with Szilard and reviewing his calculations, Einstein was quickly persuaded that the government should be warned that an atomic bomb was a possibility and that the Nazis could not be allowed to build such an unimaginably powerful weapon. On August 2, Szilard drafted the final version of the letter Einstein had agreed to send to the president. Szilard called a part-time stenographer at Columbia named Janet Coatesworth and, speaking over the telephone in his thick Hungarian accent, dictated the letter to "F. D. Roosevelt, president of the United States," advising him that "extremely powerful bombs of a new type" could now be constructed. By the time Szilard read her the signature, "Yours very truly, Albert Einstein," he was fully aware that the young woman thought he was out of his mind. That incident, no doubt exaggerated in Szilard's gleeful retelling, bears close resemblance to a passage in Richards' story in which a young secretary comes to see Perkins and confides her concerns about Zmenov. "I'm afraid he's getting himself into the most dreadful trouble," she tells him. "You know how impetuous he is. He's a genius, and when other people don't see that, he gets impatient."

Einstein's letter to Roosevelt would result in the convening of a government advisory committee to study the problem. Roosevelt appointed Lyman J. Briggs, director of the National Bureau of Standards, the government's bureaucratic physics laboratory, as chairman. On October 21, 1939, Szilard went to Washington and reported to the first meeting of the Briggs Advisory Committee on Uranium. He explained how his chain reaction theory worked and put in his usual plea for funds to conduct a large-scale experiment—the same test he had been writing to Richards about for months. To Szilard's astonishment, the committee agreed to give him $6,000 for his uranium research.

Even then, Szilard did not cease his efforts at fund-raising and kept up his letters and calls to promising prospects. Twelve days after the meeting in Washington, he sent a brief note to Richards and included an eight-page memorandum for his "personal information only," summing up his report to the Briggs committee. The memo laid out exactly how much uranium and graphite he and Fermi would need for their experiments, how much it would probably cost, and which companies could supply the materials—a blueprint for building a bomb. "It seems advisable we should talk about these things in greater detail before you take up the matter with a third person. . . ."

Szilard was never able to pin down the elusive Loomis, who a few months later would decide to back Fermi's chain reaction research. Four years later, Szilard wrote to Loomis directly, requesting an appointment to see him, and recalled his previous attempts to contact him: "I regretted very much not having been able to meet you in March and again in July of 1939 and am inclined sometimes to think that much subsequent trouble would have been avoided if a contact with you had been established at that time."

There are no records indicating whether Conant had any knowledge of Szilard's regular correspondence with Richards or his attempts to use him as a conduit to Loomis. But by the spring of 1940, when Conant found Richards' story, any public mention of atomic energy's military potential would have made the Harvard president uneasy. War had overtaken Europe, and there was already speculation about how long England would be able to fend off a German invasion. Although America was still resolutely isolationist, Conant and other leading scientific advisers to the president had been working to keep the government informed of any new developments of importance to national defense. The Briggs committee had been formed in response to the growing concern about how far along the Germans were in their atomic research. Many noted physicists, including Niels Bohr and Edward Teller and Eugene Wigner, two Hungarians now teaching in the United States, were urging their European colleagues—notably the French nuclear scientist Frédéric Joliot-Curie, the Viennese physicist Erwin Shrödinger, and the British physicist Paul Dirac—to exercise caution and were pushing for a publication ban on uranium fission. At the same time, Vannevar Bush, a tough-minded Yankee engineer who had recently resigned the vice presidency of MIT to head the Carnegie Institution in Washington, D.C., was agitating for "an accelerated defense effort." Alarmed that the United States military was technologically unprepared for war, Bush was exploring ways to mobilize the country's scientists for war.

Conant was aware that Loomis was in the thick of these talks. With close ties in the worlds of finance, government, and science, Loomis had virtually unprecedented access to the men who would ultimately decide the country's future. Not only was he a tycoon with his own advanced laboratory at his disposal, he had the financial resources to underwrite any research project he found promising, even writing a personal check

for $5,000 to help jump-start Harvard's nuclear physics research. He was
an avid supporter of leading physicist Ernest O. Lawrence and his ambitious cyclotron project—which produced radioactive isotopes that
might prove to be therapeutic or possibly provide clues to the exploita-
tion of atomic energy—and was using his wide influence among corporate chiefs and Washington officials to help Lawrence secure more than
$1 million in grant money from the Rockefeller Foundation. He was
also a first cousin of Henry Stimson, who was a member of two Republican administrations and rumored to be President Roosevelt's choice as
secretary of war. Because he had Stimson's confidence, Loomis was
uniquely positioned to play a pivotal role as the country prepared for a
war the Germans had already demonstrated would be, in Bush's words,
"a highly technical struggle."

Of course, Loomis did not need anyone's permission to undertake his
own investigation of the new machinery of war. He was enthusiastic
about American know-how and was not inclined to sit idly by until the
military, which he viewed as slow and hidebound by tradition, finally
determined it was time to take action—particularly if just catching up
with the Germans proved to be a monumental task. Long before the
government moved to enlist scientists to develop advanced weapons,
Loomis had assessed the situation and concluded it was critical that the
country be as informed as possible about which technologies would
matter in the future war. He scrapped all his experiments and turned the
Tower House into his personal civilian research project, then began recruiting the brightest minds he could find to help him take measure of
the enemy's capabilities and start working on new gadgets and devices
for defense purposes.

How much Richards actually saw and heard at the Tower House, and
how much he gleaned from Szilard or simply guessed at, is impossible to
know. What had passed for science fiction and wild speculation only a
short time ago was now no longer beyond imagining. His roman à clef
provides a rare glimpse inside Loomis' empyrean of pure science just before they would all be cast out into a corrupt and violent world. In the
final scene in his short story, Zmenov intentionally kills himself by det-
onating a small explosive "to prove forever that his theory is true."
Richards realized the race to build the bomb was on and that the coming war would change everything. He understood that the leisurely,
cloistered world of gentlemen scientists he had known at the Tower

House was at an end, and the irony that his death coincided with the passing of an era did not escape him.

Years later, Kistiakowsky's widow, Elaine, would compare Richards' stories to passages in her husband's unfinished memoir, which he had been dictating into a tape recorder up to the time of his death in December 1982. She was amazed to learn how many details Richards had drawn directly from the period the two scientists had been involved with the Tower House—from its grand beginnings in 1926 to the day it was hastily shuttered in 1940. During the decade and a half Tower House flourished, Loomis played host to a remarkable group of young scientists at a moment when new discoveries were transforming all their fields and a spirit of intellectual excitement and experimentation fueled their research. It was hard to believe that in only a few years, that bright circle would not only build the radar system that would alter the course of the war, but would go on to create a weapon that would change the world forever. "It sounds like fiction," said Elaine. "It's incredible to me now, looking back, that it really happened."

Chapter 2

BRED IN THE BONE

Ward carried himself with composure, but his politeness was merely a habit; he was preoccupied.

—WR, from *Brain Waves and Death*

"ANYONE meeting Mr. Loomis casually might find it hard to distinguish him from the great mass of men of distinction who are reared in the best families, processed by the best schools, groomed by the best tailors, and put in the vice-presidential windows of the best firms," observed *Fortune* magazine in 1946 in a flattering profile that was written with Loomis' tertiary approval but included no direct quotes or photographs of the elusive financier. "The difficulty, however, would not last long. Mr. Loomis, to be sure, has the easy and sometimes suspect charm of carefully tended manners, but in his case it is quickly evident that the manners are not the man."

Few Wall Street tycoons of his generation could match Loomis' extraordinary intellect and sheer versatility. He achieved an immense fortune in business, earned worldwide recognition for his scientific endeavors, and won the highest accolades for his service in wartime. Yet he contrived to do it all as unobtrusively as possible, choosing to remain in the background, a mysterious and remote figure. It was not that he was unduly modest so much as his wealth allowed him to do as he pleased rather than what was expected. And he had learned that anonymity served his interests.

Loomis exiled himself from the glittering world of New York society because he wanted to devote all his time to science. He set himself up royally in a castle on high hill in Tuxedo Park and financed his own audacious investigations of the stars, the heart, the brain—the secrets of the world. He built his private laboratory not as a shrine to himself, but because he desired nothing more than to be actively involved in the daily research and progress. He provided both the brains and backing for all kinds of inventions, medical advances, and scientific studies. And when duty called, he came down from his mountaintop and helped reinvent modern warfare.

"He was unique—he was primarily an unconventional person, of course, but then he was much more talented than most," said Caryl Haskins, a leading scientist who in the 1930s was inspired by Loomis, who became a close friend, to establish his own research facility, the Haskins Laboratories. "He was not motivated by money or fame. He never needed the approval of other people, he just did not need it. Well, he was that sure of himself. He was motivated purely by the facts of the case, purely by the adventure."

At the peak of his success, Loomis shunned publicity. He was secretive about his past, and colleagues knew better than to ask personal questions. "He was not so much a man of mystery as he was a very private person," recalled William Golden, a Wall Street banker who is credited with inventing an antiaircraft machine gun during World War II and later served as a science adviser to President Truman. It was in the latter capacity that he often sought out Loomis' advice. "There was a certain awesomeness about him that made him—I'm not sure what the right word is—somewhat inapproachable, I suppose. He was aloof, as if detached from a society he had once been very intensely involved in. I always got the feeling this partly had to do with his divorce, which bitterly divided his friends and family."

Loomis erected a wall between his two worlds, completely insulating his scientific Valhalla from his business life. Although he socialized with close friends from both walks of life, he never introduced a single Wall Street associate to any of his Tuxedo Park experimenters, or vice versa. "It was characteristic of Alfred that he lived in the present," said the physicist Luis Alvarez. "On the very few occasions when he shared one of the many closed chapters of his life with me, I was enchanted by what he had to say about the captains of industry and the defenders of the America's Cup, who were many years ago his most intimate friends. He apparently felt it would sound as though he were bragging if he alluded to the great power he once wielded in the financial world when in the company of a university professor."

Once, during a casual conversation early in their friendship in 1940, Alvarez happened to ask Loomis what he thought of Wendell Willkie, the Republican presidential candidate. Without hesitating, Loomis answered, "I guess I'll have to say I approve of him, because I appointed him head of Commonwealth and Southern." At the time, Loomis was the major stockholder of the power utility and was probably being nothing less than scrupulously honest in his blunt reply. But it is typical of Loomis that he regretted the slip, feeling somehow that he had let the arrogance of money leach into what he believed was a more innocent, purer part of his life. "He was immediately and obviously embarrassed by what he had said," recalled Alvarez, "and it would be another twenty years before he made another reference to his financial career in my presence."

THE aura of eccentricity Loomis acquired after he built his own laboratory, and abruptly quit the business world in favor of tinkering in his basement, has obscured how impressive and wholly respectable he was at the start of his career. Handsome, with a strong build that made him appear taller than his medium height and a massive brow framing sharply appraising brown eyes, Loomis was the very model of the bright young corporate lawyer when he entered the profession in 1912. These were comfortable times, ideal conditions for a young man to propel himself forward, and he gave every sign of having the "nicely predictable future" common to men of his background and education.

Loomis was a product of the prosperous American middle class, and though not from real wealth, he enjoyed all the same privileges, club memberships, and limitless expectations. He possessed a quiet self-confidence that was almost palpable. It was unmistakable, in the immaculate way he dressed, his calm demeanor, and the carefully modulated voice that friends and family say they never heard him raise. He took control of every conversation the same way he took control of every room he entered—he simply assumed it. Such was the force of his intelligence that most people yielded to its power. He could be ruthlessly to the point, yet his eyes would still twinkle with good humor. His daughter-in-law recalls an enigmatic charm that both men and women found compelling: "He wasn't easy," said Betty Loomis Evans, noting that beyond the standard pleasantries, he possessed no small talk whatsoever. "He was very aristocratic in a way. He was so elegant, and so good-looking, and for all his brilliance, he had a lot of warmth. He was just the best man you had ever met."

From the beginning, Loomis distinguished himself from the herd of promising young men on Wall Street by virtue of what one lawyer friend called his "Pratt & Whitney" mind, a high-powered intellect that could cut through a maze of difficulty with dazzling speed. He had a knack for making the complex appear simple, quickly visualizing a solution to a problem and laying it out before people's amazed eyes. He was relentlessly pragmatic. By force of habit, he would reduce every difficulty or decision to the remorseless logic of a mathematical equation—coolly assessing the risk factors, the odds of success or failure. "You did not want to shoot pool with him," said his friend John Foster, a former director of the Lawrence Livermore Laboratory, "because he was literally playing all the angles."

His gift was an inventive ingenuity, an almost childlike ability to look at something as if for the first time and take it apart and re-create it along the lines of his own imagination. More often than not, he also had the attention span of a child. Curious, eager for the next astounding and unanticipated result, he would no sooner notch a contribution in one endeavor than he would take off in an entirely new direction. He attacked new problems with a single-minded zeal, rattling off "ninety ideas per minute," according to one associate. He loved puzzles and never tired of trying to work out explanations for life's riddles—the smallest increments of time, prime numbers, black holes—with anyone who could follow his lightning-quick reasoning. He excelled at games of all kinds. By age nine, he was a chess prodigy and regularly amazed his peers at St. Matthew's Military Academy in Tarrytown, New York. By thirteen, he could play "mental chess" without aid of a board or pieces and could play blindfolded, carrying on two games simultaneously. He impressed his classmates at Phillips Academy, Andover, when, on a school outing to a New York City chamber of horrors known as the Eden Musee, he challenged a costumed character named the Masked Marvel to a game of chess and played him to a draw.

The first sign of "precocity," according to family members, showed in his early facility for magic. From the time he was a small boy, Alfred was fascinated by the art of illusion and would not rest until he could work out the solution and re-create it himself. He became a master of sleight of hand, making quarters appear and disappear behind ears and inside pockets, and could perform thrilling card tricks. He collected the apparatus used by professional magicians and staged elaborate shows for his siblings and cousins. His younger sister, Julia, once recalled, "You never knew if you were on the ceiling or on the floor when Alfred was around." He never lost his touch. Throughout his life he would demonstrate a talent for the surprising, inexplicable, remarkable result.

"He liked to awe people," recalled his grandson Alfred Lee (Chip) Loomis III. "Magic fits into that psychological profile. He was a very imperious type. The whole idea of having your own laboratory and inviting people to come there and live like kings and do science—it was a pretty extraordinary arrangement. The whole concept has to do with exercising a sort of subtle power and control."

L O O M I S ' brand of ingenuity cannot be taught, it is bred in the bone. It was also peculiarly American. By background and temperament, he was well prepared to respond to the momentous changes taking place in the twentieth century and to exert his influence over events in the farthest reaches of the social, financial, and scientific worlds. His lineage was colonial on both sides, but in his forebears' questing spirit and intellect, he was a child of the future.

From his earliest memories, there was never a time Loomis was not aware that he bore an exalted name, synonymous with a life of great success, service, and honor. It was an indelible part of his identity. His grandfather Alfred Lebbeus Loomis was a tuberculosis specialist who earned worldwide recognition for his advances in the treatment of pulmonary diseases and was one of the most honored doctors of his generation. He was also an educator, reformer, and leading philanthropist who was elected president of virtually every prestigious medical society. Regarded as one of New York's first citizens, he lived in a big house at 19 West 34th Street and, despite giving away large sums of money, managed to leave an estate worth in excess of $1 million. If Alfred Loomis would later strike some of his peers as princely in his bearing, it was in no small part because of his grandfather's august achievements and his deeply ingrained belief that the rich should repay their debt to society. Throughout his life, Alfred Loomis would feel that moneymaking alone was not a satisfactory existence.

The Loomises came from a long line of solid, resourceful Yankees. Alfred Lebbeus Loomis was the seventh-generation descendant of Joseph Loomis, a woolen draper from Essex, England, who sailed from London to Boston on the ship *Susan and Ellen* in 1638 and settled with his wife, Mary White, on a twenty-one-acre plantation in Windsor, Connecticut, the site of the first English colony in the state. He was born in Bennington, Vermont, in 1831. There had been much tuberculosis in his family, and he developed an early interest in medicine because of his own weak lungs. As a child, he was convinced he "would not live to be over thirty." But after a sojourn in the Adirondacks restored his health, he became persuaded of the curative powers of the mountain air and devoted himself to the study of respiratory problems. He spent at least two months of every year in the Adirondacks and became an enthusiastic champion of preserving the region's forests as a "natural sanitarium."

After graduating from the College of Physicians and Surgeons, Alfred Lebbeus Loomis became a practicing physician in New York, treating consumptives in the charity hospitals on Wards and Blackwell's Islands. He was appointed to the staff at Bellevue and, later, Mount Sinai Hospital. He was active in establishing a home for consumptives in Saranac, New York, while another, in Liberty, New York (later renamed the Loomis Sanitarium), was one of the first to provide poor invalids with a level of care that had previously been available only to the rich. In his later years, he channeled his considerable charisma and passion into fund-raising, becoming a tireless campaigner for medical research. He personally donated $15,000 to the University of the City of New York and persuaded the local grandees to give much more. In 1886, he announced a gift of $100,000 to the medical school from an unnamed wealthy "friend" to build and equip a new laboratory.

After his death in 1895, the university's chancellor, Reverend Henry MacCracken, noted in his memorial address that "the generous founder insisted on but two conditions: that it should be called the Loomis Laboratory, and that the giver's name should remain unannounced. . . .

"Dr. Loomis not only got money from the rich for good objects," said MacCracken, "but he won the hearts and minds of men to something else than merely spending their incomes."

A tall, well-made man with "a roaring voice and contagious laugh," the good doctor earned a reputation among his fraternity of physicians as an epicurean. He was much toasted and celebrated after his will revealed a $10,000 gift to the Academy of Medicine to establish "the Loomis Entertainment Fund," the income to be used for the supper at the monthly meetings. A gleeful posthumous tribute in the *New York Herald Tribune* recalled that "as master of the feasts," Alfred Lebbeus Loomis excelled: "He was an authority on canvas-back duck, knew all the best vintages by their first names, and could tell from what region in Cuba a perfecto came from watching the smoke."

He had two children, Henry Patterson Loomis and Adeline Eliza Loomis, who were born and educated in New York City, schooled in his views on health, science, and good citizenship, and expected to make their contribution to society. After their mother, Sarah Jane (Patterson), died in 1880, he married Anne Prince, a wealthy widow. They all participated in his work at the Loomis Sanitarium, and his son, as expected, followed him into medicine. After graduating from Princeton and New York University Medical College, Henry Loomis went abroad to study for a number of years. In 1887, he returned to New York and took up his father's work—joining the staff of Bellevue, continuing his research in diseases of the heart and lung, and seeking election to the same medical boards, societies, and clubs. He also took up his father's convivial ways, only more so.

Henry Loomis was twenty-eight years old and Julia Stimson twenty-six when they married later that same year. He was tall and handsome and regarded as one of the most dapper young doctors in town. She was the proud, striking daughter of a distinguished American family, one of seven children of the Wall Street banker Henry Clark Stimson and Julia Atterbury Stimson. Henry Clark Stimson had been "one of the most respected men in the market," but he ended up losing most of his fortune and, after the Panic of 1873, retired from business. For the next twenty years, he lived quietly on his small savings and the income from his wife's trust. Julia Stimson was raised in the shadow of his disappointment in an austere bourgeois household on East 34th Street, just a few minutes' walk from the Loomis family home. Her marriage to the prosperous young doctor, son of an esteemed healer and philanthropist, must have seemed like a fortuitous match. In what was no doubt interpreted as a good omen, their first son was promptly ushered into the world nine months later.

Alfred Lee Loomis was born on November 4, 1887, and named for his grandfather, although they changed the biblical Lebbeus to Lee on the grounds that it sounded "too old-fashioned." Two more children followed: a daughter, Julia Atterbury Loomis, four years later, followed by another son, Henry, who was named after both his father and uncle, Henry Stimson. But it was not a happy union, and Alfred's mother, whom the children called Belle Mere, would endure the humiliation of hearing rumors of her husband's charming way with female patients. She would return often to the Stimson home just down the street, where she could always count on finding solace and support. As the marriage deteriorated, the couple formally separated, and rumors of their pending divorce scandalized New York's Victorian drawing rooms. "You could live apart, that was not at all uncommon," explained Betty Loomis Evans. "But they were actually going to divorce, which was *never* done. They were practically ostracized from society."

In those days, divorce befell a respectable family like a tragedy, and the sense of shame was grimly communicated to the children. Alfred's sister, Julia, never forgot the social censure they endured. Years later, she would still talk about the way she was snubbed by "good families" because of the cloud of disrepute hanging over their heads. During her debutante season, when she was forced to sit out some of the parties, her older brother made it his business to introduce her to eligible young men. "Her father was quite a ladies' man," said her youngest son, Ed Thorne. "He just walked out on the family. My mother never referred to him, but it was always clear that she was very bitter."

"Julia always said they had a pretty rotten childhood," added Evans. "I think she suffered much more, or at least more openly, than he did. Alfred and Julia were very, very close growing up, and he always felt his first duty was to make sure his sister was taken care of. He was terribly protective of her. Alfred was determined Julia would never want for anything."

The two had always been inseparable, bound together from early childhood by the shocking death of their little brother. It had happened before their eyes, just as the family was preparing to set off on an outing. Alfred and Julia were already seated in the horse-drawn carriage, and their baby brother, Henry, was about to be lifted in when a rabid dog rounded the corner and attacked him. The toddler was badly mauled before the dog could be pulled off him. He was rushed to the doctor, but there was no cure for rabies in those days, and he could not be saved. It was an awful death. Alfred, who was almost ten at the time, never forgot

Henry's terrible screams and convulsions. "Alfred watched his little brother die right there, right in front of him," said Bart Loomis, who can recall his grandfather's vivid retelling of the incident. "The suddenness of his death taught him something that day he never forgot, and it formed the bedrock of his determination. He knew what it meant to take care of his family. It was seared into him."

A regular escape was eastern Long Island, where his Stimson aunts rented houses each summer, taking along with them their many nephews and nieces. Alfred and Julia spent much of their time with their large band of Stimson cousins. As they grew older, summers were divided between their mother's place in Sterlington, New York, near Tuxedo Park, and Highhold, the rambling, thirty-acre Stimson home-stead in Huntington, Long Island, that belonged to her brother, Henry. The family was presided over by their maternal uncle, Lewis Atterbury Stimson, a powerful, commanding personality whose intelligence was supported by a "rugged, militant character." Lewis Stimson had served with distinction in the Civil War and went on to become an accom-plished doctor and one of the first American surgeons to operate using antiseptics. But he never recovered from the tragic death of his wife nine years after their marriage, and he withdrew into the workaholic grind of emergency medicine, abandoning his son, Henry Stimson, and a daughter, Candace, to be raised by his aging parents and an unmarried sister, known as Aunt Minnie. A stoic, somewhat removed figure, he re-mained the undisputed patriarch of the "fifty uncles, aunts, and cousins" that constituted the extended Stimson clan and were the abiding influ-ence of Alfred Loomis' life.

It was within this family circle—bonded by unhappiness, orphaned children, absent parents, and interconnected households—that Alfred Loomis and his cousin Henry Stimson, twenty years his senior, formed their extraordinarily close, lifelong allegiance. When Henry Stimson discovered early in his marriage that he had been left sterile by an adult case of mumps, his young cousin became the son he would never have. Alfred Loomis idolized Henry Stimson, and Henry became the surro-gate father from whom he inherited his single-minded purpose, as well as a dominant characteristic that was famously known by members of the clan as "the Stimson reserve."

The men in the Stimson family were encouraged to turn to profes-sions rather than to business—one of Henry's brothers was a clergyman, another a lawyer—marking a strong intellectual strain that was fairly unusual in well-to-do American families. Most of the important decisions of Loomis' early life proceeded from the collective wisdom of the Stimson men, who were consulted at every turn and weighed in with their expectations, imperatives, and stern injunctions. Central to the Stimson way of thinking was membership in the elite, which in their view was accomplished by attending the right institutions. So after a stint at St. Matthew's Military Academy in Tarrytown, it was determined that Loomis should follow the same path as Henry Stimson—first Andover, then Yale, then Harvard Law School.

Alfred was a gifted, if somewhat distracted, student. While at Andover, he reportedly "burnt up" courses in mathematics and science but demonstrated only passing interest and grades in the required language and arts courses. Throughout his schooling, he was more interested in his hobbies and inventions than he was in the drudgery of daily assignments. But Andover provided a welcome refuge from the tensions at home, and he would later remember his time there with affection. He was captain of the chess team and the tennis team and was voted "Brightest" in the class his senior year. Like many young men of his age, he was obsessed with cars and had the pocket money to indulge his hobby. His dormitory room was infamous for its phalanx of wired and radio-controlled cars and trucks, which could mow down an unsuspecting visitor. Writing to Henry Stimson in April 1905, Loomis' mother conveys a clear understanding of her son's extracurricular enthusiasms:

> I had a letter from Alfred yesterday in good spirits, in which he asked
> me to "throw on third gear" and write immediately. His mind is evidently occupied with the delights of motoring to come. As for the
> present every opportunity is given him to concentrate upon study. He
> writes "it appears to be clear sailing to a diploma."

When he entered Yale that fall, Loomis had not yet settled on any one course of study and chose not to enroll in the Sheffield Scientific School. In his freshman year, he made something of a name for himself when one of his devices went slightly awry. Loomis had boldly announced that he had a contraption that would enable him to "hypnotize anybody" in a matter of minutes. He had with him a small black velvet box containing a crystal, and when a small crank on the side was turned, it flashed a light on the sparkling stone, presumably producing a hypnotic effect. One student volunteered to be his next subject and was led away by Loomis to a dormitory room for the experiment. In just two turns of the crank, he passed out. Unfortunately, Loomis' triumph was short-lived when the student could not be revived, and a doctor had to be hastily summoned. It turned out that the boy was in an "epileptic coma" and the hypnotism was purely coincidental, rendering inconclusive the results of his most famous experiment to date.

The correspondence between Loomis and Henry Stimson reveals that as a college student he continued to consult his increasingly eminent cousin on all matters large and small, whether it was requesting permission to keep his automobile in New Haven so he could "take it all apart and overhaul it" or asking for help on a paper about the power of the president. Stimson, who served as President Theodore Roosevelt's U.S. attorney for the southern district of the state of New York from 1906 to 1909, always answered promptly and respectfully, establishing a lifelong exchange of information, advice, and ideas. Forwarding Loomis a document he had written the previous year on the president's authority in regard to the army, Stimson wrote: "It was a brief in defense of his discharge of the members of the Twenty-Fifth Colored Infantry after the riot in Brownsville, Texas. You may find something in it which may interest you. I should be glad if you send it back when you are through."

At Yale, Loomis majored in mathematics and quickly distinguished himself as a brilliant, if highly original, thinker. During his junior year, he took a postgraduate course in advanced calculus taught by Ernest William Brown and attended by a select few, including a Chinese student and a lady mathematics teacher. It soon emerged that Loomis and the professor were the only two people who had any idea what was going on in the class: "The textbooks were in German and French, and a typical day's assignment was ten problems," according to E. Farrar Bateson, a corporate lawyer and lifelong friend. "There was a regular ritual each day. Brown would enter the classroom and say, 'Good morning, ladies and gentlemen, did anybody get any answers?' The only one who invariably had all the answers was Alfred, and his answers invariably disagreed with those of the professor. Brown would look them over and say, 'Those are very good answers. The only trouble with them is that they are incorrect.'"

His interest in physics first revealed itself in his study of boomerangs. At one point, Loomis and Brown were observed all around campus throwing boomerangs, "mysteriously writing figures on their cuffs." It turned out they were trying to get the contraptions to conform to a theoretical formula for boomerang flight that Loomis had worked out. At the time, Alfred was intrigued by the aeronautical advances being made by the Wright brothers, and he carried out his own experiments with kites and gliders. He designed a model airplane based on a theory he had evolved. The plane resembled a "venetian blind," and according to his theory, the layered wings would fan out in flight and yield a greater combined lifting action than the standard model. However, when he hooked up his model to a car with a kite line and took off, it "promptly plummeted to earth." A glider he constructed and tested on the beach near the Stimson home in Long Island fared better, staying aloft several minutes. He served as class secretary his senior year but reportedly devoted the bulk of his time and attention to his technical pursuits. He was known to have an unrivaled ability to bone up on a new interest in almost no time at all, and one of these self-taught areas of specialization—a fascination with artillery weapons and a huge store of arcane data on their country of origin, design, and capabilities—would later prove most useful.

Loomis was a sophomore at Yale when his father died quite suddenly, after a five-day bout of pneumonia. He was only forty-eight years old, but he had long ago ceased to be the major influence in his son's life. It is a reflection of the bitter state of affairs that family members would often say of his father that "at least he had the good grace to die before the divorce." While Loomis had given little thought to a career up to that time, except that he would probably engage in some kind of scientific work, he laid aside those dreams and set out on a career in the law, fully aware that the responsibility to support and protect his family now fell to him. He enrolled in Harvard Law School and threw himself into his coursework and became an editor of the *Harvard Law Review*. He graduated cum laude in 1912 among the top ten in his class. Contemplating his future after Harvard, Loomis, as was his habit, checked in with Henry Stimson. "My dear Cousin Harry," he wrote, confirming his intention to pay him a visit in Washington over the Thanksgiving holiday: "My plans for the coming year are very uncertain, and before making any decisions I want to have a talk with you, if possible."

After passing the bar examination, Loomis started as a law clerk in Henry Stimson's firm of Winthrop & Stimson, obtained entirely "on his merits," as George Roberts, a senior partner in the firm and also a product of Yale and Harvard Law School, would always insist. Henry Stimson was not there to welcome him personally, as he was still in Washington, serving a two-year stint as secretary of war in the Taft administration. But the entire firm had about it the feeling of extended family and was staffed by young lawyers whose fathers were either related or friends, and who overwhelmingly shared the Yale/Harvard imprimatur and, as Stimson put it, "the fellowship of club members."

The chambers at 32 Liberty Street (then 32 Nassau Street) occupied a suite of spacious offices on the fourteenth floor of the Liberty Mutual Building, with sweeping views of the East River and the Brooklyn Bridge. The firm maintained the comfortable, ordered atmosphere of eighteenth-century London chambers, with the clerical staff gathered in a large central room, which opened onto the high-ceilinged library, where law clerks researched cases within earshot of the partners' private offices. It was still in the days when clerks got their training in the general practice of law by working closely with partners on many different kinds of cases, and Loomis was assigned to corporate and financial work and soon showed a flair for security issues, mortgages, and reorganizations. He quickly earned a reputation as one of the more outstanding young lawyers on Wall Street and in only three years was made a member of the firm.

Winthrop & Stimson was a distinguished firm and held it as a point of honor, as Stimson once explained it, that they "stood outside the Wall Street group" and did not "adopt the methods of the others." The result was that they also did not get as much business as the others. In the years leading up to World War I, the senior partners at Winthrop & Stimson were making a respectable $20,000 a year, while lawyers in a neighboring firm were earning five and ten times that amount. Stimson, who returned to the practice of law in 1913, and his partner, Bronson Winthrop, were of the old school and believed the day began at nine-thirty and ended at five-thirty—except when a case was in court—and discouraged work on Saturdays or discussing cases after dinner. This meant that the firm regularly turned away clients who were seeking the kind of high-powered counsel they could call upon night and day, whose law clerks could be relied upon to slave into the early morning hours. Stimson was so determined to discourage certain kinds of clients that after hearing out a group of men from the West Coast who had a scheme to evade the antitrust laws, he turned them away with the statement "I can just hear the gates of the jail clanking shut behind you."

That Stimson was able to build up a very lucrative practice despite his "New England conscience" proved just how able an attorney he was.

So it was in such an impeccable, if somewhat poky, firm that Loomis began his career, with plenty of leisure time on weekends to devote to a wife and family. At twenty-four, he married Ellen Holman Farnsworth, two years his junior, a delicate beauty from a distinguished Boston Brahmin family and the sister of a Harvard classmate, Henry Weston Farnsworth. The young Miss Farnsworth was reputed to be "the prettiest girl in Boston" and a great catch. She was typical of her day and social class: intelligent, educated—she could read French, Latin, and Greek—and utterly impractical. Tutored at home, she had been immersed in music, poetry, and opera since childhood and had a heightened commitment to the life of the mind. Nervous and high-strung, she was given to terrible headaches that would leave her prostrate for days. She believed in "dressing for dinner" and never descended in anything but the most feminine concoctions cascading with lace. They made a stunning couple, though she struck some observers as extremely old-fashioned and an odd choice for her energetic, forward-looking husband. But Ellen was devoted to fulfilling his every want, respected his desire for large amounts of solitude, and diligently schooled herself in his arcane scientific interests. They were married on June 22, 1912, at her family's country home in Dedham, Massachusetts, and honeymooned in England, staying in a castle fully staffed with servants that was loaned to them by a family friend. "Alfred is very happy," Loomis' mother reported in a hastily scrawled note to Stimson.

The couple first made their home in Sterlington, where his mother already owned a home, and eventually moved to Tuxedo Park. For the next few years, Loomis focused on his career and family and blended in with so-called young marrieds who had settled in Tuxedo. They played tennis and golf, attended genteel soirees, and were praised in the social columns. The couple had two sons in quick succession, Alfred Lee Loomis Jr., known as Lee, and William Farnsworth (Farney) Loomis. "Although their names will not appear in the Social Register for some time, there are several important newcomers in the fashionable set," society columnist Cholly Knickerbocker announced in the *New York American*.

But after only three years, the marriage was blighted by tragedy. At the end of September 1915, Alfred and Ellen were rocked by the news that her brother, who had impetuously enlisted in the Foreign Legion at the beginning of the year, had been killed in the bloody battle for the Fortin de Navarin in Champagne. Farnsworth was every bit as bookish and idealistic as his sister and, after graduating from Harvard, had notched some experience in the Balkans as a reporter for *Collier's* and the *Providence Journal* and published a book chronicling his adventures, *The Log of a Would-be War Correspondent*. He had sailed for the Continent again the previous fall and after only a few months got caught up in the military fever that was sweeping London and Paris; in January he joined the Foreign Legion. Henry Farnsworth was the first of the young American volunteers killed in World War I.

Loomis was unprepared for the depth of his wife's sadness. She had been extraordinarily close to her brother and had passionately supported his noble impulse to enlist. When he had gone abroad in 1912, Ellen had beseeched her husband to ask Henry Stimson, who was in the middle of a two-year term as secretary of war under President William Taft, to arrange a letter of introduction for him from the State Department to the United States diplomatic officers. She had eagerly awaited his vivid dispatches describing the life of the legionnaires. For months after his death, she was inconsolable. She wept for hours at a time and withdrew into her own world, writing long, grief-stricken letters to her mother every day. "She spent a lot of her early married days in mourning for him," recalled Betty Loomis Evans. Years later, Ellen was still moved to tears by a passing reference to Farnsworth. "She was devastated by his death. She and her brother were so full of glorified ideas about knights in shining armor. They both had this romantic idea of war. I don't think either of them was too connected to reality."

WHEN the United States declared war against Germany in April 1917, the twenty-nine-year-old Loomis promptly enrolled in officers training camp and was sent to Plattsburgh, where his mathematical skills quickly became apparent. Artillery work came easily to him, and he amazed his fellow officers with his knowledge of modern field artillery, an obsession dating back to his college days. Loomis had spent the months before the war boning up on his old hobby. Along with a handful of other young lawyers, he had organized the "Canoneer's Post" to study the latest ordnance equipment being mobilized by the European powers, and he had persuaded West Point instructors to act as their tutors. As a result of his expertise, he was commissioned a captain on July 16 and transferred to the army proving ground in Sandy Hook, New Jersey.

Stimson naturally oversaw Loomis' wartime postings as closely as he had his civilian career. For Stimson, the war was a moral duty, and although fifty years old and almost blind in one eye owing to an old injury, he insisted on active duty and was eventually appointed a lieutenant colonel in the field artillery. But he believed Loomis' talents would be best served behind the lines and urged him to apply for a job at the new Aberdeen Proving Ground in Maryland under General William Crozier, one of "the most able and progressive officers in the army." Stimson forwarded Loomis' application to Crozier, along with a formal letter of recommendation and a personal note outlining his abilities. In a letter to Loomis that spring, Stimson counseled his young cousin not to worry that the Aberdeen assignment would "necessarily deprive you of the opportunity for field service." He added:

> I find no one who differs with our view that your most promising avenue for usefulness lies along this line. I have great confidence in your inventiveness and thorough grasp of the principles of mechanics. It is quite possible that, with the aid of these qualities, you might, by a single solution of any one of the many mechanical problems which are confronting us in this war, accomplish more than could be accomplished by five hundred line officers in the artillery.

On January 1, 1918, after four months at Sandy Hook, Loomis was sent to Aberdeen. His "inventiveness," along with a constant stream of ingenious ideas for new armaments and new solutions to old tactical problems, had earned him one of the most important jobs: chief of the development and experimental department. At Aberdeen, he joined some of the most distinguished physicists and astronomers in the country, who had been mobilized to help adapt the army to the needs of modern warfare. One of Loomis' responsibilities was to test ideas for new weapons submitted to the Army Ordnance Board. Among the luminaries whose novel recommendations Loomis had to take into account and duly try out were those of Thomas Edison, who was by this time a genuine folk hero, his life story and list of major inventions—the stock ticker, the incandescent electric light bulb, the phonograph, the moving picture, the alkaline storage battery, and a host of others—known to nearly every schoolchild. Although he was already seventy years old and crustier than ever, he was still hard at work in his laboratory in Orange, New Jersey, and throughout World War I contributed technical advice and suggestions.

Loomis followed up on one recommendation and recalled that the "perilous experiment" not only blew up Edison's scheme, it almost annihilated a number of Aberdeen personnel as well. As the story was later recounted in *Fortune* magazine, "Edison's idea was to work up a terrific spin on a large TNT-filled drum, and then suddenly drop it from a height. The theory was that the drum—trailing wires attached to a detonator—would bounce along of its own momentum until it reached the enemy lines, when it could be exploded. Put to the test, the drum dug a hole in the ground where it fell, bounced out, traveled about 200 feet, dug another hole, and then let go."

The hazardous duty must have been noted, because shortly afterward Loomis was promoted to the rank of major and put in charge of experimental research on exterior ballistics. "He was the military officer in charge of the small R and D division," recalled Paul Klopsteg, an electrical engineer, "when I came there to take on research and development, principally on methods for measuring projectile velocities. So he and I worked together rather closely during those days, and then after we moved to Aberdeen and continued there." At the time, there was no simple way to measure the velocity of shells fired from guns, making it difficult to predict with any great accuracy the time it would take for a shell to arrive at a given point. The existing method, called the Boulenge chronograph, was extremely complicated, difficult to operate, and only fairly accurate. It measured the time it took for a shell to pass through two wire screens, the recording mechanism activated by the projectile as it passed through the screens and broke the electric circuits. It also had a cumbersome custom-made design that was impossible to mass-produce, making it impractical for the modern military.

Loomis was convinced the device could be improved. Working with Klopsteg, he developed an ingenious new method for measuring the velocity of shells. One of the problems with the old "break circuit" chronographs was that the wire could stretch until it reached the breaking point, maintaining the circuit for varying lengths of time depending on the angle of the shell at impact and creating irregularities of some magnitude. Loomis and Klopsteg devised an electric recording instrument that, for the first time, depended on closing rather than opening the circuits, thereby eliminating the chief source of error. It consisted of an aluminum disk spooled with ticker tape, which a small motor kept revolving at a constant speed. Instead of breaking a circuit on contact, the shell created an electrical impulse when it hit each screen, causing a spark to burn a small hole in the tape. Shell velocity could easily be calculated by measuring the distance between the spark holes.

The revolutionary new device, known as the "Loomis chronograph" and later formally called the "Aberdeen chronograph," was simpler to use and far more reliable. It was also admirably adapted to field use because of its portability and immediately supplanted its predecessor. "In its present form, it is the result of the enthusiastic initiative of Major A. L. Loomis," trumpeted an article in *Army Ordnance*, "and the ability of Dr. Paul E. Klopsteg to cope with the various physical and electrical problems in the development of such an instrument." Hundreds of Aberdeen chronographs were made during the war, and as the result of electronic improvements added in World War II, Loomis' and Klopsteg's remarkably efficient invention became standard U.S. Army and Navy equipment.

Loomis was promoted to lieutenant colonel and up to the close of the war worked on a variety of ideas that would later materialize as military advances—from a recoilless cannon to a low-slung French 75 hung on a tripod mount for concealment purposes that was dubbed the "snake in the grass." But the Aberdeen chronograph was his first invention—his first scientific triumph. He took enormous pride in his accomplishment: his name was listed first on the patent filed by the United States Army. Not long afterward, "Alfred L. Loomis of Tuxedo Park" applied and received a second patent of his own, "for an improvement in chronographs" design to enhance the construction, and thus the accuracy, of the rotating drum within the recording device. "Even though he was just a young lawyer," said Luis Alvarez, "with his mathematical background, and his wonderful intellectual talents, it turned out that he really did a good job and invented a number of things that were used for twenty-five years in measuring external ballistics."

Loomis remembered his time at Aberdeen fondly and loved to tell amusing stories about his experiences with the antiquated ways of the United States Army. When he was first assigned to work with an outfit of cannoneers, he recalled being puzzled by one of the company's soldiers, who always walked fifty paces or so in back of them and would stand stock-still for hours at a time with one arm slightly raised. When

Loomis finally asked what on earth the man was doing back there all by himself, he was astonished to learn that he was filling a role that dated back to when cannons were still pulled by horses. One soldier always led the horse and held the reins, and even though the horses were long gone, the post had never been abolished. Loomis formed a lasting impression of the military bureaucracy as fundamentally averse to change, and that prejudice would never leave him and would influence much of what he did in the next war.

He also became an enthusiastic champion of the new armored tanks, and at every opportunity lectured Stimson that they were the future of modern warfare, "not horses and guns." He became such an expert on tank construction, he built a scaled-down model in his garage at Tuxedo Park in order to see if he could make further improvements in the design. On one memorable occasion when Colonel Stimson came to visit, Loomis rolled into Tuxedo's small Victorian rail station in his light armored tank to meet the train, kicking up dust and causing quite a scene. When they had both climbed on board and were noisily clanking home, Loomis turned to Stimson and said with satisfaction, "Now this is the way to protect the nation."

WHILE at Aberdeen, Loomis struck up a friendship with Robert W. Wood of Johns Hopkins University, widely considered to be the most brilliant American experimental physicist of his day. Loomis had met Wood before, owing to the slight acquaintance of their families, both of which had summered in eastern Long Island. The two men hit it off, bound by the instant kinship two scientists can discover when they share a playful intellect and a passion for the same discipline. Wood was from a wealthy New England clan, like Loomis the precocious son of a doctor, and from childhood he too had had an absorbing interest in all sorts of scientific phenomena. He had twice flunked out of the Roxbury Latin School before being admitted to Harvard, where he earned a bachelor's degree in chemistry. He took a leisurely approach to graduate work at Johns Hopkins, until he was distracted by what was going on in the physics laboratory next door. He had little patience with the formalities of academe and never bothered to complete his Ph.D., which no doubt only further endeared him to Loomis.

Handsome, charming, and eccentric, the fifty-year-old Wood had earned worldwide fame as the investigator of scientific frauds, undertaking highly unusual inquiries—for a serious scientist—that led to his becoming quite well-known to the public. In 1904, Wood had been instrumental in discrediting the French physicist Rene Blondlot, who claimed to have discovered a new source of radiation, and had published an article in *Nature*, one of the leading international scientific journals, that exposed the N-ray delusion that had swept physics laboratories in France. He was also known for helping to unmask several leading spiritual mediums as outright fakes. He had no patience for scientific quackery and set ingenious "bear traps" for those who claimed to have made some secret invention that they were actually using to con investors. Shortly after the war, in perhaps his most famous escapade, he aided the New York police in the "Wall Street bomb case," in which a big barrel bomb was exploded in front of the Morgan Bank minutes before noon, killing thirty-nine people, crippling scores of others, and wounding more than four hundred pedestrians. Wood managed to reconstruct the bomb from only a few fragments found on the scene, which resulted in his being dubbed by the popular press as the "Sherlock Holmes" of science.

While Wood was more scientist than inventor, he was not solely concerned with academic research. If he saw a practical application, he would pursue it but he rarely reaped any financial benefits from his findings because he left them to others to follow up and develop. Along the way, Wood was responsible for a number of significant inventions, including various applications of invisible ultraviolet radiation. He was one of the first to make successful photographs in infrared and ultraviolet light and applied this technique to the detection of art forgeries. He also used it to photograph the surface of the moon and planets in order to discover details invisible to the eye and was credited with a number of other advances in photography. He developed what is known as "Wood's glass," an efficient filter that, combined with a mercury lamp, suppresses the visible light while freely transmitting the ultraviolet. During the First World War, the "invisible light" was used for signaling without alerting the enemy. He also discovered that many substances fluoresce brightly when subjected to ultraviolet light and immediately spotted its potential for stage performances. Florenz Ziegfeld happened to be Wood's East Hampton neighbor, and not surprisingly, any number of his optical effects found their way to the *Ziegfeld Follies*.

But for all his stunts and showmanship, Wood was a brilliant and creative experimenter. During his brief tenure at the University of Wisconsin, he invented what became the standard way to thaw frozen underground water pipes by passing an electrical current through them. He moved to Johns Hopkins in 1901 and remained there for the rest of his career. Like Loomis, he was a compulsive tinkerer and inventor whose curiosity knew no bounds. That Loomis was a lifelong gadgeteer and largely self-taught naturally appealed to Wood. Before long, the friendship evolved to the point where the flamboyant and unconventional Wood, almost two decades older, had become Loomis' mentor and, in effect, role model.

By the spring of 1919 the war was over, but Loomis' scientific education at the hands of *The Amazing Doctor Wood*—the modest title Wood later assigned to his biography, which he was rumored to have more or less dictated—was just beginning. Loomis returned to Winthrop & Stimson but only lasted a few months before his restlessness, combined with a redoubled interest in science and inventions, mandated that he make a fresh start. He was determined to escape the drudgery of law and to find a way of financing his interest in science. According to Luis Alvarez, it was axiomatic that for Loomis life would never be the same again: "He had had a wonderful taste of experimental physics," observed the 1968 Nobel Prize winner in physics, whose first exposure to the subject during his junior year at the University of Chicago he compared to falling in love at first sight. "So he set up his own laboratory at Tuxedo Park where he lived."

Chapter 3

THE POWER BROKER

Ward was not a man to do things by halves.

—WR, from *Brain Waves and Death*

LOOMIS returned from the war determined to make a fortune, and he looked no further than his brother-in-law, Landon Thorne, for his new partner. Landon Ketchum Thorne had been a year behind Loomis at Yale. After graduating from Yale's Sheffield Scientific School in 1910, he had made a good start selling bonds for the Central Trust Company, the predecessor of Hanover Trust Company. He soon jumped to the old, respected investment house of Bonbright & Company as a junior salesman. Loomis thought Thorne was a comer and had introduced him to his sister, Julia, and could not have been happier when the two were married at the small church in Tuxedo Park in 1911.

At first glance, it seemed an unlikely alliance, with the gregarious, backslapping Thorne—a popular college athlete who had basked in the reflected glory of his cousin Brink Thorne's Yale football career—almost in every way Loomis' mirror opposite. But Thorne recognized his brother-in-law's mathematical genius and its application to the financial game. He exploited what he later referred to as "the general unsettlement of thinking after the war" to talk Loomis into throwing over a promising legal career for the much more speculative investment banking business. "He somehow persuaded Alfred, who was the fair-haired young man at Winthrop Stimson at the time," recalled his son, Edwin Thorne. "But my father was a good salesman, and he persuaded him to give up the law and join him. They were not good friends particularly, but they obviously knew each other, and Father had a great admiration for Alfred's talents. It was an interesting jump, and an interesting combination of two people."

"Landon was an Irish charmer," said Evans. "He was smart, and so handsome, and full of such joie de vivre. He could sell anything to anybody. No one could say no to him, not even Alfred."

Thorne had his own reasons for wanting to go into business with his brother-in-law. By the end of the First World War, Thorne, who had served as an army captain during the war, was also feeling restless upon his return to Bonbright, which had seen better times. In the intervening years, most of the firm's founding members had died, retired, or lost interest, and the brash, energetic young Thorne had made quite a name for himself, in one year selling so many bonds that his commissions ran up into the six figures—more than the partners were pulling down in profits. At thirty, he was already widely regarded as one of Wall Street's most resourceful and coolheaded operators. As a result, the powers at Bonbright huddled and invited him in as a partner. Thorne naturally accepted, only to discover that the partners had been kept in the dark about the company's precarious financial health and the business was on the brink of ruin. The postwar financial slump had dried up a lot of the firm's old business and left it with a vast assortment of unrealizable assets, not the least of which was a firm that owned prunes in now Bolshevist Russia.

After reviewing the company's figures with Loomis, who had become well acquainted with the house's affairs while serving as its legal counsel at Winthrop & Stimson, Thorne was convinced it might be possible to revive Bonbright's flagging fortunes. He proposed a plan to Loomis in which the two of them, with the pledged support of Charles A. Coffin, president of General Electric, would form a partnership to quietly buy up majority control of Bonbright. Together, they could reorganize the firm and set off on a more lucrative course. According to Edwin Thorne, who followed his father and older brother into the family business, "Father figured he had to get somebody really knowledgeable, with a scientific turn of mind, because he wanted the firm to move into financing utility companies—he figured that was the area to be in."

And Loomis had all the right contacts. At Aberdeen, he had met many of the top scientists at General Electric who were developing new things for the utility business, including bigger transmission lines and other modernizations. Loomis was quickly convinced of the partnership's potential and wasted no time breaking the news to Stimson. Characteristically, his cousin was equally prompt in his reply. He sent Loomis a telegram expressing his hopes that they could still meet as planned, concluding simply, "ROBERTS [senior partner at Winthrop & Stimson] WILLING TO RELEASE YOU."

Loomis returned to his old law firm only long enough to clear out his desk. By the fall of 1919, he had joined Thorne at Bonbright's offices at 25 Naussau Street. The two brothers-in-law pooled their resources and scrounged capital from relatives. Thorne asked several family members, including his father, to back them, but only his uncle, Samuel Thorne, came through with the money. Then one day in 1920, they broke the news to their partners that the investment firm they had organized in secret, Thorne, Loomis & Co., had bought a majority share of Bonbright. There followed a series of sudden retirements, and the founders, William P. and Irving W. Bonbright, liquidated their interests. Loomis and Thorne took control of the company, and the bloodless coup— which came to be known as "the deal by which the Bonbrights vanished from Bonbright"—sealed the young upstarts' cutthroat reputation. Almost at once, they began specializing in public utility issues and quickly emerged as leaders in the financing and developing of the electric power industry.

What appealed to Loomis was the challenge of shaping the nascent industry. He had no respect for the old school Wall Street capitalists' skills. He relished the opportunity to reinvent their creaky methods and along the way rewrite the rules as he saw fit. Rural electrification was the future, the key to the growth of new factories, industries, and economic opportunities. Loomis had complete confidence in the new technology as a force for change and a force for good. If they could speed the growth of the power industry, both Bonbright—and the country— would benefit.

The brothers-in-law surprised even themselves with how well they worked together, combining Thorne's sense of what his business needed with Loomis' sense of how to deliver it. Little time passed before Loomis and Thorne grew to be close colleagues and friends, discussing business deals, supporting each other's best efforts, and sharing their weekend enthusiasms for golf, tennis, fishing, and sailing. Thorne preferred the outdoor life and made the long commute every day to his Wall Street office from a large estate in Bay Shore, Long Island. He always carried a supply of chocolate bars in his pockets to sustain him on the journey and was known for passing them around when negotiations ran late. Thorne was the partnership's public face and driving deal maker, while Loomis provided the ingenuity and imagination that made it possible to pull off what would later come to be regarded as textbook transactions. "Landon was the outside guy," said his grandson Landon Thorne III, a lawyer and businessman in Beaufort, South Carolina. "Alfred sat there with the slide rule and figured out the bond yields, then Grandpa went out and sold them."

A banker who worked with them reported: "Loomis had ninety ideas a minute; Thorne knew how to pick the good ones and put them to work."

In the 1920s, public utility companies were growing rapidly, and Loomis and Thorne were among the first to recognize the need for credit to finance the huge expansion ahead. The percentage of American homes wired for electricity had increased from only 8 percent in 1902, to 24 percent in 1917, to more than 34 percent by 1920. The industrial demand for electricity had also increased exponentially as factories shifted from steam to electrically powered machinery and equipment. The rate of technical advance was tremendous, and individual operating companies were cropping up across the country, expanding electrical service to the public and materially reducing costs. The furious rate of expansion had created a bottleneck, with many operating companies unable to issue bonds and secure investment.

Bonbright led the way, reorganizing and liquidating old-fashioned companies formed by capitalists who had little knowledge of the new industry and replacing them with new ones. Loomis and Thorne helped create utility holding companies by bundling the management and facilities of smaller operators into larger integrated systems, or "superpowers." This allowed the operating companies to obtain funds by issuing securities and thereby enlarge their operations. The holding companies were also a better medium for investors, who instead of taking common shares in one holding company were able to invest in a diversified group. Loomis and Thorne also understood early on that the only way to provide low-cost power to great population centers, as well as to serve large numbers of rural communities, was to finance and build power plants. They perfected the use of the holding company as an instrument for raising capital. Along the way, they researched and explained the intricacies of utility regulation to an increasingly interested press, public, and Congress, no doubt occasionally interpreting the rules to their advantage.

Loomis and Thorne were willing to underwrite security issues most of the old-line houses shied away from. When the boom in the utility market came along, they were perfectly positioned to reap the rewards: between 1924 and 1929, Bonbright, either alone or with associates, did upward of $1.6 billion worth of utility financing, underwriting roughly 15 percent of all the securities issued in the United States. Their clientele included most of the big American utility systems, and they played a major role in organizing and consolidating interconnected utilities into the two largest of the so-called superpowers: the American Superpower Corp. in 1923 and the United Corp. in 1929. As a result, they earned board seats and unprecedented influence in an enviable stable of companies, including Public Service of New Jersey, Electric Bond & Share, Consolidated Gas of New York, Niagara Hudson, and Commonwealth & Southern, among others.

Loomis and Thorne's phenomenal nine-year run, reviving Bonbright from near bankruptcy to its vaunted status as the leading private investment house specializing in utilities, became a Wall Street legend. They were corporate capitalists of the first order, yet they were lauded not only for their spectacular success, but for not being motivated solely by a concern for private industry and for applying scientific principles and long-term economic planning to the management of public resources to provide cheaper, more reliable service to consumers. Together, they conceived and promoted the concept of the big holding company— which along with many other ideas was later adopted by the Securities and Exchange Commission—and were in large part credited with the friendly relations among the major eastern power systems. Their advice was sought after by foreign governments, and in 1928 they helped the Italian government set up and manage the Italian Superpower Corp., based on American Superpower, the model of the modern public utility. At the time, it was hoped that the influx of American investment would help speed the development of Italy's power industry, reduce its coal imports, and help strengthen its faltering economy.

Fortune described their achievement in a gushing profile in the premiere issue of the magazine in February 1930: "Bonbright has become not only the industry's banker, not only its spokesman, but to some extent its guide. Thus it is that some utility men consider Messrs. Thorne and Loomis, Bonbright's president and senior vice president, as the most potent force in shaping the present and future organization of America's huge, complex power and light business."

IN a matter of only a few short years, Loomis and Thorne had become both powerful and very prosperous. But Loomis was never at peace with himself about spending his days preoccupied with money. He had come to believe science was a higher calling and was troubled by the growing sense that much of his work was self-serving and profited men he neither liked nor much admired. He could not forget the pride in achievement he felt at Aberdeen, culminating in the award of the chronograph patent. Even in the midst of exhaustively analyzing and preparing deals for Bonbright, he was always absorbed by some new discovery or advance in research he had stumbled across in his reading.

As he had done since boyhood, whenever he found some small problem that intrigued and puzzled him, he would devote hours in the evening and on weekends to clear up the matter. He usually devised some simple solution that went straight to the heart of the problem, and was forever toying with some new invention or gadget that he believed more efficient. Throughout his early years at Bonbright, he repeatedly badgered a patent attorney, Robert Byerly, to apply for copyright registrations on various small "devices" he had invented, ranging from designs for his own slide rule for calculating securities to drawings for a "simple, reliable fire extinguisher." After the U.S. Patent and Trademark Office rejected his customized slide rule, Loomis wrote dejectedly to Byerly in September 1922: "I don't believe it is worthwhile to take out a patent on these slide rules. I appreciate very much the interest you have taken in this little device, and if you will send me your bill for the copyright and the time and attention that you have given to this matter I will have a check sent right over." But Loomis persevered and eventually obtained a patent for his new and improved fire extinguisher, which replaced the then common pump model with a closed receptacle that contained the fire extinguisher liquid under constant pressure, as he put it, "so that it may be projected upon a fire by the mere opening of a valve."

Loomis kept in close touch with many of the physicists he had met at

Aberdeen and followed the new research they were doing. On a summer's day in 1924, when Loomis was visiting his Stimson aunts in East Hampton, he decided to call on Wood, who had a farmhouse nearby that he kept as a summer retreat. Loomis found Wood at work in his barn laboratory behind the house, and they had a long chat about postwar research and swapped stories about everything and anything they had seen or heard of "science in warfare." After that, Loomis got in the habit of dropping by to talk almost every afternoon, Wood later recalled, "evidently finding the atmosphere of the old barn more interesting if less refreshing than that of the beach and the country club."

It was during one of these leisurely afternoon sessions that Loomis asked if there was any research Wood could contemplate the two of them doing together, perhaps an area "which required more money than the budget of the Physics Department could supply." Loomis added that he "would like to underwrite it."

Wood had become interested in physical optics, especially spectral studies. In the barn, he was busy putting the finishing touches on a forty-foot grating spectrograph—then the largest and most powerful in existence and capable of giving better results than anyone had ever seen before. While it appeared crude, in Wood's capable hands it performed superbly, and Loomis was transfixed. As unconventional as Wood's methods were, the results were always near perfection. For example, his preferred method for cleaning the giant spectrograph—which when unused for any extended period became clogged with spiderwebs—was to simply drop the family cat in one end. In order to escape, the poor animal had to make its way through the whole length of tubing, effectively running a fur duster through the works.

Wood had a keen eye for social opportunities, and like most scientists, he was always short of funding for his research. He knew of Loomis' wartime interest in Professor Paul Langevin's research in Paris in the field of supersonics. Langevin had developed a method for detecting submarines by sweeping the sea with a narrow beam of high-frequency sound waves and picking up the echo or reflected sound with special electrical apparatus. During the war, Wood had gotten himself assigned to Langevin and had gone with him to the naval arsenal at Toulon, where his apparatus was in operation. He saw that when fish swam across Langevin's beam of high-frequency sound waves, they turned belly up and died, and when a hand was held in front of it in the water, there was a painful burning sensation.

Wood outlined an elaborate course of research he knew Loomis would find impossible to resist: "I told him about Langevin's experiments with supersonics during the war and the killing of fish at the Toulon Arsenal. It offered a wide field for research in physics, chemistry, and biology, as Langevin had studied only the high frequency waves as a means of submarine detection." Wood suggested that he and Loomis continue Langevin's work and investigate some of the biological and chemical effects of high-frequency sound. In return, he knew he could count on Loomis' generous support of his own research in optics.

Loomis, of course, was impatient to get under way. The two arranged a trip to General Electric's research laboratory to purchase the high-powered vacuum tube oscillator they would require for their research. Loomis returned home and began making plans to equip an improvised laboratory behind his house in Tuxedo Park. After the birth of his third son, Henry, Loomis had purchased a much larger Tudor home on Club House Road in Tuxedo Park, which was designed by the fashionable Philadelphia architect Wilson Eyre. The stately brown stucco mansion was nestled on the side of a steep incline, with three acres of lush gardens and rolling lawn spreading out behind it. Just down the road, at the bottom of the hill, was an old barn that served as a garage.

It was there that Wood proceeded to train his wealthy protégé in his methods and in the process achieved some amazing results, beginning with the development and redesign of a far more powerful model of the General Electric oscillator. It was originally built in Schenectady, and together Loomis and Wood dismantled it and then installed their reconfigured model in the largest room in Loomis' garage. Wood described their first project:

> The generator was an imposing affair. There were two huge Pliotron tubes of two kilowatts output, a huge bank of oil condensers, and a variable condenser with intersecting wings of the type familiar to every amateur radio operator, but about six feet high and two feet in diameter. Then there were the induction coil for stepping up the voltage and the circular quartz plate with its electrodes in an oil bath in a shallow glass dish. With this we generated an oscillatory electric potential of 50,000 volts at a frequency of from 200,000 to 500,000 per second. The oscillating voltage applied to the electrodes on the quartz plate caused it to expand and contract at the same frequency, and generate supersonic waves in the oil. . . .

Using their pliotron oscillator, similar to the high-frequency oscillators then used in radio broadcasting, and "stepping up" the voltage from the usual two thousand watts up to fifty thousand watts, Loomis and Wood were able to produce "super–sound waves" with a frequency of over two hundred thousand per second—more than ten times beyond what is audible to the human ear. To protect their eardrums from damage, they plugged their ears with cotton and donned earmuffs for protection while performing their research. When they passed the oscillating electric current through a natural quartz crystal, it caused the crystal to expand and contract sometimes up to half a million times a second—which could hurl the quartz to pieces. To prevent this, they immersed the quartz in oil, which absorbed the vibrations and rose up in the form of an erupting volcano two inches high, producing a fountain of oil drops that shot a foot or more into the air.

Experimenting with their super–sound waves, they took a small glass tube and drew one end of it out into a thread of glass as thin as a hair. The other end they dipped into the oil, and the tube was shaken by the vibrations. When Loomis held the thread lightly between his fingers, he felt no trace of the turbulence caused by the waves. But when he pressed the thread tightly, it burned a deep groove in his skin. They discovered that that same thread could burn through a piece of wood and bore a hole through a plate of glass like a power drill. Transmitted through water, sound waves of this frequency caused the water to churn furiously as though boiling, but with little increase in temperature, kicking up a disturbance, or "pulse wave," very like the cloud of white spray that used to arrive on the surface of the ocean after a depth charge was exploded.

But that was only the beginning. Upon investigation, they found the sound waves produced other strange chemical tricks: applied to the contents of a test tube, they shook two fluids immediately into a mixture, even breaking down mercury into such fine particles that it hung suspended in water and formed an emulsion. They brought about the so-called clock reaction, in which a transparent solution quickly changes to dark blue, and also appeared to be able to change paraffin into a crystalline form.

The work Wood and Loomis were doing at Tuxedo was garnering attention in scientific journals in the United States and abroad, in no small part due to the sensational way Wood had of presenting his scientific data. He had P. T. Barnum's touch for creating the excitement and publicity that guaranteed that his talks were always well attended. The day before he was due to read a paper on his experiments on sound waves before the National Academy of Sciences in Washington on April 26, 1926, *The New York Times* trumpeted the first of many breakthroughs at Tuxedo Park: "Dr. R. W. Wood, Professor of Experimental Physics at Johns Hopkins University, today made public the results so far attained in the experiments conducted on the estate of Alfred L. Loomis, a New York banker, at Tuxedo, N.Y., with treatment of diseases by high frequency sound waves sent through water. Mr. Loomis assisted in the experiments."

The *Times* went on to explain that the possibility of applying the discovery to medicine lay in stimulating circulation, which could be achieved in any part of the body that was submerged into water in which the sound waves were then introduced. The discovery was thought to be particularly helpful in the treatment of arthritis or gout. Some physicians believed that arthritis was caused by organisms that left chalky deposits on the bone at the joints, and it was thought that the sound waves might stimulate circulation and help carry off the deposits. "Dr. Wood said that while the experiments had not gone far enough for him to claim the cures might be accomplished, it had been found that a method for stimulating circulation without injury was valuable to medicine."

As the scope of their research expanded, they became pressed for room in the garage, and Loomis began hunting for a larger building in which he could house his workshop. By now, Pierre Lorillard's planned resort—"a short season place between New York and Newport"— which had welcomed three trainloads of fashionable visitors in a gala opening celebration in 1886, had swelled to a prosperous suburban community of the social elite. After selling his grand Newport mansion, "the Breakers," to the Vanderbilts, Lorillard had wanted to simplify his life and commune with nature. He had envisioned Tuxedo Park as an isolated community carved out of the rocky cleft and had hired the architect Bruce Price to build a colony of rustic, shingle-style "cottages" that would blend in with the beautiful wilderness setting. Since the beginning of the century, when the upper classes began indulging in transatlantic travel and seeing firsthand the castles of the European aristocracy, a more ostentatious style had been in vogue. Over the next two decades, Tuxedo Park had become a greenbelt of extravagant mansions, the bigger the better. Many of them were monuments to vanity—and folly—and Loomis had his choice of architectural extravaganzas, ranging from Tudor, Gothic Revival, Spanish Mission, Georgian, Jacobean, and Queen Anne.

In 1926, after a survey of available properties, Loomis settled on the Tower House, as it was known to locals, a huge, crumbling mock Tudor pile of stone and masonry built on a high thrust of land lying between Crow's Nest and Ant Hill Roads, on the farthest southeastern tip of Tuxedo Park. Of course, "huge" was a relative term in Tuxedo, and Loomis' acquisition was dwarfed by many neighboring properties, including Henry Poor's gargantuan redbrick mansion and the sprawling Brook Farm, which belonged to the chairman of the National City Bank. Tower House was modest by comparison, a large three-story stone-and-timbered mansion constructed on solid bedrock, complete with a steep gable and crenellated stone tower, patterned brick chimneys, and elaborate stained-glass and leaded diamond-pane windows. Though it had been derelict for more than a decade, the house had a craggy charm and a lodgelike simplicity that appealed to Loomis. Everything about the place, from its name to the panoramic views of the wooded hills and shimmering lake below, spoke of history, legend, and myth—what better foundation for a great laboratory?

The house had originally been built by Spencer Trask, a prosperous investment banker, for a rumored $500,000 at the turn of the century, when Tuxedo first flourished as a fashionable resort for millionaires. But after a series of long illnesses and the death of their firstborn child, Spencer Trask and his wife, Katrina, a poet, decided to move to Saratoga Springs, New York—it was said that Katrina Trask could not bear to return to the house where she had borne so much unhappiness—where they built an even more magnificent mansion they called Yaddo, after a word their daughter made up to rhyme with shadow.

The Trasks' Tuxedo home took its name from its subsequent owner, one Joseph Tuckerman Tower, a New York businessman who bought the property in 1907 and immediately began extensive renovations. Unfortunately, Tower's improvements extended to blasting the vast cellars of his new home, apparently in an ill-advised attempt to rid the place of field mice. "[He] must have been a somewhat bizarre character because he had a phobia about mice," said Paul Klopsteg, who was a frequent visitor and heard the details of the house's strange history from Loomis. "He had all the ground floors made of reinforced concrete so that the mice couldn't get in."

As the mice would not be deterred, Tower blasted out a new basement subsequent to the house's construction. He kept up a steady drum-roll of explosions until his blue-blooded neighbors gathered rank and, in the name of the almighty Tuxedo Park Association, ordered him to stop. And the Tuxedo Park Association, vigilant watchdog of the park's territorial integrity and social purity, was not to be ignored. Tower was so outraged by this public censure that he stormed out of the house and, as legend has it, "called for his carriage and four, and drove off—never to return." After his death, the property became a frozen asset of his estate. For years, the Tower House stood empty and abandoned. Its solitary tower and weather-beaten battlements barely visible above the heavy mists, the wind whistling through the broken stained-glass windows and sixteen master bedrooms, the Tower House gave rise to all sorts of ghostly tales, passed down to Tuxedo children by their nurses.

When Loomis bought the unmanageable white elephant in 1926—picking it up for $50,000 or, as he later put it, "for a song"—local Tuxedoites just shook their heads. But even Loomis, who thoroughly enjoyed frightening his boys with tales about the haunted house, was momentarily spooked by the sight that greeted him when he unlocked the doors and stepped inside. Tower had apparently been in such a hurry to close up the house, he must have ordered the workmen to drop what they were doing and leave at once, herding them out the door himself. When Loomis walked into the house for the first time, said Klopsteg, "there were all the lunches the workmen had brought in." The mice had eaten the last crumb, and all that remained were the shredded sacks and scattered cups and knives.

As it turned out, the concrete floors were ideal for scientific instruments, so Loomis converted the cavernous basement into an enormous elaborately equipped experimental room. Much of the first floor, including some twenty servants' quarters, pantries, and storerooms, was turned into office space. He even installed a machine shop complete with a full-time mechanic. Loomis restored the impressive exterior of the house, refurbished the stately upstairs bedrooms and lounges, and preserved the chapel and splendid grand salon and ballroom with their dark wood paneling and heavy furniture. He also left the small theater, which Tower had reportedly built for his wife, who liked to put on dramatic performances.

When Wood arrived to inspect their new headquarters, he described the building as almost palatial, "a huge stone mansion with a tower, like an English country house, perched on the summit of one of the foothills of the Ramapo Mountains in Tuxedo Park. This he transformed into a private laboratory de luxe, with rooms for guests or collaborators." Loomis had moved Wood's forty-foot spectrograph from the barn in East Hampton and installed it in the basement, as Wood put it, "so I could continue my spectroscopic work in a better environment." But to Wood's surprise, Loomis had taken the liberty of making a few minor adjustments on his famous invention:

> [He] had a new tube made for the instrument, since there was no point in digging up the underground sewer pipes which had served formerly. He packed the tube in boiler felt with an arrangement for keeping the entire tube at a constant temperature, had a new and better camera made, installed motors, revolution counters, etc., for rotating the grating, which was housed in a small closet built around the brick pier on which it was mounted, and arranged other substitutions and gadgets, until I told him there was nothing left of my celebrated spectrograph but the forty feet. It had experienced a "reincarnation," and required no pussycat as housemaid.

In the meantime, Loomis was eager to meet some of the celebrated European physicists and visit their laboratories. He asked Wood if he would go abroad with him. In the summer of 1926, the two men set off on a grand scientific tour of Europe, which they would follow up with a second trip two years later.

They sailed for England on the *Île de France* in early July and were met at Plymouth by a Daimler, in which they were driven to Hereford for a visit with Wood's friend Thomas R. Merton, a professor of physics at Oxford and later treasurer of the Royal Society. The visit turned out to be particularly memorable because his estate bordered on the river Wye, and their arrival coincided with the salmon fishing season. As Wood recounted in his biography: "Merton had a fine private laboratory behind the house and some interesting experiments to show, but for once Loomis was excited over something other than physics. He waded in the Wye and landed a fifteen-pound Salmon."

They continued on to Paris, where they toured a number of "superb laboratories," including that of Dr. Jean Saidman, who was investigating the medical applications of ultraviolet light. Saidman was enthusiastic about Lumiere Wood, the name the French had given his wartime invention, and proudly offered to give them a demonstration of his state-of-the-art X-ray machine with a fluoroscope. Loomis leapt at the chance to witness firsthand the workings of the human stomach. So with Wood acting as the guinea pig, and after a dose of barium carbonate, Loomis got his wish—with Wood holding a mirror so that he could "witness the process too."

On the voyage home on the *Olympic,* Loomis began making ambitious plans for a princely private laboratory of his own. He was tremendously excited by what he had seen in Europe. He took particular interest in Merton's estate, which followed in the tradition of Terling Place, the famed residence of Lord Rayleigh, born John William Strutt, the English physical scientist who had won the Nobel Prize in 1904 for his successful isolation of argon, an inert atmospheric gas. Rayleigh had constructed his laboratory adjacent to his manor house, and it was there that he had carried out practically all of his major scientific investigations. Loomis now had a blueprint for the kind of research facility he envisioned at Tuxedo Park, where he hoped to do the kind of significant scientific work accomplished in the first Loomis Laboratory built by his grandfather. Never one to doubt himself, he purchased an enormous leather guest book with gilded pages in which to record the names of all the famous scientists who would make the pilgrimage to the Tower House in the years to come. The first name, signed in flowing black ink, was that of Robert Williams Wood, whom Loomis appointed as the laboratory's eminent "director of research."

Thrilled by their initial success, Loomis was eager to continue where they had left off, and he and Wood moved on to a series of experiments to explore the effect of their discovery of supersonic sound waves on living organisms. For this they called in Dr. E. Newton Harvey, professor of biology at Princeton and a national authority on living cells. Looking through a high-powered microscope, Harvey saw that the waves had the faculty of breaking down blood vessels and would therefore have a deadly effect on small animals and fish eggs and even on certain kinds of plant life. Spirogyra, or the familiar bright green scum found on pond surfaces around Tuxedo Park, died and vanished completely after being exposed to the supersonic rays for five and one half minutes. Over the next two years, recalled Wood, they experimented with the supersonic "death rays": "We worked together, killing fish and mice, and trying to find out how and why they were killed, that is whether the waves destroyed tissue or acted on the nerves or what."

In January 1927, the *New York American* carried the headline "Super-

Rays Discovery of Rich Banker." The story provided a breathless account of their work: "Alfred L. Loomis, Wall Street banker and scientist, with a beautifully equipped research laboratory at his Tuxedo Park residence where he cooperates in pure science necromancy with his friend, Professor R. W. Wood, of Johns Hopkins University, made public yesterday details of a new form of sorcery—super-audible sounds. . . ." In his first public interview as a scientist, Loomis, who was more cautious than Wood, tried not to overplay their discovery: "We cannot tell yet, of course, where this will lead to in the future. Our discoveries with these super-audible sounds may bring about highly valuable results in biological research . . . what we have discovered will no doubt be useful to science, but not to warfare or to medicine in the treatment as disease, as far as we now know."

In September 1927, Loomis and Wood published their findings, which they entitled "Communication No. 1 from the Alfred Lee Loomis Laboratory," in the highly respected British *Philosophical Magazine and Journal of Science*. Their classic paper "The Physical and Biological Effects of High-Frequency Sound-Waves of Great Intensity" inspired related studies that kept researchers all over the world busy for years. They also received a patent for their methods and apparatus for forming emulsions through powerful high-frequency compression waves in liquid. That same year, working with William Richards, who was then a young chemistry professor at Princeton, Loomis published a second paper, "The Chemical Effects of High Frequency Sound Waves."

Loomis wanted to be published in scientific journals, primarily because he felt that other research scientists should know about his findings and would naturally be as fascinated as he was. At Wood's suggestion, he often submitted his scientific papers simultaneously to journals in the United States, Britain, and the Continent, which saw to it that word of the Loomis Laboratory's research studies would quickly spread.

In the coming years the Loomis Laboratory would publish twenty more papers on the effects of high-intensity sound waves: Loomis was a coauthor of the first four and, several years later, another four papers. Together, he and Wood did the first major work in the field now commonly known as ultrasound, and their names still appear in some textbooks as the "fathers of ultrasonics." The field has since grown enormously, and the applications extend from industrial cleaning and emulsifying to medical imaging, with ultrasonic scanners now in use in hospitals to observe fetuses, watch the motion of heart valves, and detect tumors.

By the autumn of 1927, Wood's own research in spectral studies was opening up new fields for study. For the past several months, he had been absorbed in the study of fluorescence in various gases. He had found that by shining a light of a specific wavelength through a gas, he could obtain a series of bands called a "resonance spectrum." The spectrum varied as a function of the wavelength of the illuminating light. The results of his experiments had profound implications for theories of the atomic structure of matter being debated at the time, particularly Niels Bohr's theory. Wood's discoveries in spectroscopy led to his being nominated that year for a Nobel Prize in physics. While the prize went to Arthur Compton, Wood's research was attracting prominent scientists to the Loomis Laboratory; in all, ten papers in optical spectroscopy would come out of the laboratory, including one by Loomis and George Kistiakowsky. But spectroscopy was relatively well-trodden ground, and Loomis set out to do pioneering work in new fields.

Loomis continued in his capacity as vice president of Bonbright five days a week, devoting nights, weekends, and vacations to an expanding array of research experiments under way at Tower House. He was now thoroughly committed to his second career as a scientist and piled extra work on himself, spending every spare hour reading the journals and books Wood recommended as part of his ongoing education. "He used to drive into Wall Street every day and do his business, and then come back to Tuxedo Park where he had this marvelous laboratory," said Alvarez. "In fact, he had a much better laboratory than any university laboratory at that time—better equipment, more expensive equipment. He hired R. W. Wood as his private tutor, and Wood came up and spent every summer at Tuxedo Park doing the experiments that he couldn't do at Johns Hopkins because they didn't have enough money. R. W. Wood taught Alfred Loomis physics."

The single-minded zeal with which Loomis pursued his research meant that he absented himself from the house on Club House Road. Repeating the Yankee tradition in which he was reared, he packed all three of his boys off to boarding school from the age of seven: first to the Fay School, then to St. Paul's, and finally to Harvard. Ellen Loomis had been more or less always "unwell" for years, and Loomis saw to it that she underwent the latest experimental "radiation treatments," though they seemed to do her no good (and may well have done her real harm).

Ellen admired and supported her husband's passion for research, and while she took pleasure in arranging the elaborate dinners and soirees at Tower House, she increasingly felt an outsider to his new world. At the Tower House, he had set up an independent household removed from Club House Road, where women were tolerated but not welcome. In a letter to Stimson in August 1927, Ellen apologized for being unable to make their annual visit to Highhold, Stimson's Long Island estate, and described how much life had changed at Tuxedo Park:

> Alfred's work in the laboratory fills me with pride—just thinking of doing it seems to me rather wonderful—just the business part of planning and arranging that house, and starting it, has been quite a performance. But his work has blossomed out, the results of his years of lonely study are showing so suddenly, it is very thrilling just to be me, and watch. It is even more thrilling to be Alfred, and do. It has required a quality of fierce absorption that has shut out many important things from his life temporarily. I think it is worth it.

Determined to be cheerful, Ellen wrote she was confident that she had found "a number of ways in which I can help him, as housekeeper, and hostess, and doer of odd jobs." She hoped they would manage to get away the following year, but at present they were too distracted by the publication of Loomis' first scientific papers, including the one with Wood in the *Philosophical Magazine* and another, written with E. Newton Harvey of Princeton, on the biological effects of high-frequency sound waves, in an upcoming issue of *Nature*. She urged the Stimsons to come to them in Tuxedo for a few days so they could show him "the Tower." She added reassuringly that she was certain calm would soon be restored: "This white heat of Alfred's work is somewhat due to the extra things that come of a new venture. A certain amount of routine will develop presently, and that is always easier."

The following winter, at Wood's suggestion, Loomis hosted "a congress of physicists" designed to be the new laboratory's official coming-out party. The conference, "Certain Aspects of Atomic Physics," was in honor of James Franck, professor of physics at Göttingen University, who along with Gustav Hertz had won the Nobel Prize in 1925 for their work on the laws governing the transfer of energy between molecules. Franck had agreed to give his first lecture in the United States at Loomis' laboratory, and dozens of leading scientists traveled to Tuxedo

Park for the event. Papers were also presented by Wood and a young MIT physicist named Karl Compton, brother of the recent Nobel Prize winner, who had struck up a close friendship with the unconventional financier. Held in the library at Tower House, a room of "cathedral-like proportions with stained-glass windows," the conference proved such a success that they immediately made plans for another one the following year.

Loomis discovered he enjoyed the role of host, and in his own autocratic style, he began regularly holding large house parties, complete with honored guests, designated lectures, music recitals—often featuring virtuoso performances by one of the visiting scientists—and lavish formal dinners with a liveried servant standing behind each chair. "I gave a series of weekends, thirty and forty people were invited guests," Loomis later recalled. "When Bohr first came over to this country he'd give a series of talks, what we now call seminars. In those cases, all kinds of people came. Marconi came over and came out. For a time it got to be if any distinguished European scientists came to America, Tuxedo was so near, they were met at the boat and came out." Loomis' private laboratory was rapidly acquiring world fame as a center of research. It had truly become, in Einstein's phrase, a "palace of science."

Chapter 4

PALACE OF SCIENCE

> A group of men were seated in a straggling half-circle in the
> large upper room which served as lounge, game room, and
> occasionally as auditorium. Broad foreheads, sparkling
> spectacles, and one van Dyke beard gave the gathering an
> appearance which was convincingly scientific.
>
> —WR, from *Brain Waves and Death*

B Y the summer of 1928, Loomis was the subject of frequent and sensational press reports. In his gusto to put his Tuxedo laboratory on the map, he had become more of a public figure than he had anticipated, and he found he lacked Wood's taste for the limelight. Newsmen had fastened on to the Tower House as a story and Loomis as an eccentric, would-be Einstein—the "fantastic dreamer incarnate"—and were having a field day writing about his ghoulish experiments with the silent sound waves they dubbed "the whisper of death." He had wanted only to join the priesthood of the nation's men of research, and his quest for recognition had opened the gates of the temple for all manner of undignified and unseemly inspection.

In the 1920s, science was enjoying a tremendous popular resurgence, and the burgeoning mass-circulation press, aided by the advertising industry, had become propagandists for the advances of modern technology, daily trumpeting such marvels as Einstein's "revolutionary" theory of relativity—locked in the atom, reported the *Saturday Evening Post*, was "a source of power inconceivably greater than any possible requirement of the human race"—to the latest high-powered vacuum cleaner. Einstein was front-page news, and reporters followed his every move, documenting his self-effacing mannerisms and utterances as further evidence of his genius. He was "the world's most celebrated scientist," noted the historian Daniel Kevles, and his cult status "not only helped enlarge the prestige of pure science," it endowed the entire profession with a kind of awesome glamor. By 1925, the *New Republic* wrote that scientists were regarded as members of an exclusive and powerful fraternity: "Today [the scientist] sits in the seats of the mighty. He is the president of great universities, the chairman of semi-official government councils, the trusted adviser of states and even corporations."

If Loomis found his new high profile uncomfortable, there was little he could do to restrain the tabloids. Every time he gave a lecture or presented his research to a gathering of his peers, the local press picked up on it and ran an overblown account of the wizard of Tuxedo Park, "an expert performer of sleight-of-hand tricks," and his "magical" laboratory. He soon became such a fixture of the columns, *Popular Science Monthly* devoted a breathless feature story to his double life, called "A Scientist of Wall Street":

It is the peak of a rush day on the New York stock market. In the office of the vice president of a large Wall Street banking house, a dynamic, boyish-looking man sits at the throttle of a high-speed machine of finance. About him seethe the hubbub and excitement of the world's money market. Quick decisions, hurrying messengers, the steady grind and chatter of the stock ticker—each moment is crowded with feverish activity.

A few hours later, on a broad estate at Tuxedo Park, N.Y., miles from the city frenzy, this same high-powered business executive may be seen hard at play. In white apron, surrounded by curious test tubes, chemicals, and electrical apparatus, he is taking his recreation—in a physics laboratory!

The man is Alfred Lee Loomis, physicist, business man of science. The laboratory is his private playground. . . . The Loomis Laboratory is known to scientists the world over. For, from his playtime research in collaboration with technical men of note, who accept his hospitality and the use of his fine equipment, have come some of the newest marvels of discovery in physics and biology. . . . Through a high-powered microscope, Loomis has seen waves of silent sound twist and shatter human blood corpuscles into a thousand bits, or hurl cells of living protoplasm into a mad, whirling dance of death.

A fascinating hobby, this—to turn from the rumble of Wall Street to play with "whispers of death."

Loomis declined to be interviewed for the article. He was deeply embarrassed by the emphasis on his financial success and, more to the point, worried that the story might be interpreted by his scientific colleagues as self-aggrandizing at a time when Einsteinian humility was the academic uniform. He was also concerned that any undue credit attributed to either himself or his laboratory might alienate the serious researchers he was courting for future projects. He did not want the carefully calibrated team play at Tower House disrupted by frivolous publicity. "Loomis, hard-fisted man of affairs, is reticent about his laboratory," the magazine noted in the last paragraph of the story. "He likes best to pursue his pastime with his friends, without the public eye upon him."

Inevitably, the Tower House was becoming the focus of all kinds of strange rumors. Tuxedoites took note of the odd comings and goings at the big house on the hill, and the ill will it engendered was tinged with anti-Semitism: "Strange outlanders with flowing hair and baggy trousers were settling down for weeks and months on end. They were performing all kinds of crazy experiments—cooking eggs and killing frogs with sounds that nobody could hear, clocking time to the ten-thousandth of a second, making turtles' hearts beat in a dish, and similar enormities."

There had been a number of large gatherings of scientists, and a veritable parade of foreigners from all over the world had been allowed past the massive stone walls of Gate Lodge, the imposing, isolating guardhouse that Lorillard's architect Bruce Price designed to protect the gated colony from trespassers and the public at large. Known as "the world with a fence around it" for the eight-foot-high barbed-wire fence that had originally encircled the colony to keep out the riffraff, Tuxedo had always been famously inhospitable to outsiders, especially Jews. "Woe to the unlucky stranger who strays across the posted boundaries," warned the New York Herald. Leaving no room for error, Lorillard had spelled out the restricted membership list of his club as "a guide to Who is especially Who in the Four Hundred." Once the exclusive stomping ground of Astors, Juilliards, Goelets, Tuckermans, and Pells, oldtime Tuxedoites took a dim view of Loomis' imports and complained there was now no telling who would arrive on the five o'clock special from New York. Perhaps as a precaution against this attitude, Loomis often privately reserved the famous Tuxedo Club Car and arranged to have Erie trains make unscheduled stops at Sterlington, where his driver and Rolls could regularly be seen waiting to whisk his guests up to the Tower.

Kistiakowsky, who was a frequent visitor at this time along with Richards, always had the uncomfortable feeling he was trespassing: "It was a strictly WASP community, minorities of all kinds not being welcomed," he wrote in his memoirs. "Many square miles of fenced and patrolled hilly terrain, a large private lake with wonderful sailing, a golf course and so on, including lovely rocky forest scenery, owned by quite a selection from the 'New York 400' families. . . ."

At the broad-verandahed Tuxedo Club, where members gathered to drink and play cards, it was whispered that Bonbright's brilliant vice president was neglecting his Wall Street business "to putter around at all hours in the laboratory." The clubhouse was the center of life in the park—and, in the minds of many residents, the world—and some perceived Loomis' self-imposed absence as a slight. After all, many newly minted millionaires would kill to make the rosters of the storied club, which had helped establish the popularity of court tennis, racquets, and golf—it boasted the second oldest course in the country—as well as the abbreviated dinner jacket called the "tuxedo." (Legend has it that Griswold Lorillard, Pierre's son, once attended a ball in a dinner jacket without tails, a style he had copied from the Prince of Wales, and it was adopted as the club's informal uniform.) Loomis and his wife put in an occasional appearance at Tuxedo functions, and their three sons actively partook of club life, but he made it a rule never to mix science with society. They received friends and neighbors at their home on Club House Road and reserved the Tower House exclusively for their learned guests.

While most Tuxedoites, who came from banking and railroad fortunes, relaxed with a round of golf or a game of court tennis, Loomis "was all business all the time," recalled John Modder, who worked as a bellhop, waiter, and bartender in those days. "You'd never see him down at the club on a Saturday. He was always up in his laboratory." Modder was on friendly terms with the butler at the Tower House and occasionally had a chance to visit the lab. "Now I used to laugh, because he had a great big painting on the wall, and at the bottom of the painting he had a pool with fish in it, like an aquarium. You could see the reflection from the painting in the water. And Mr. Loomis would say, 'You look in the water. There's a dead lady in the bottom of that lake.' Well, I looked for hours, but I couldn't find it."

"We knew Alfred Loomis had bought the old Tower House and was doing experiments up there," said John Jay Mortimer, whose father, Stanley Grafton Mortimer, was a member of one of Tuxedo's founding families and whose mother, Katherine Tilford Mortimer, the daughter of Standard Oil founder Henry Morgan Tilford, was one of the leading arbiters of park society. "People thought him pretty eccentric. Absolutely nuts, really."

But even if his more narrow-minded neighbors had wanted to stop him, it is unlikely they would have tried, let alone succeeded. Loomis' research had important medical and therapeutic applications and had won the respect of scholars from Johns Hopkins, Princeton, and Harvard. He was also one of them—he "belonged" in Tuxedo society, as did his mother, whose social credentials were impressively attested to when she was elected president of the Colonial Dames of America and the exclusive Monday Opera Club (the latter was limited to two hundred members). And while the Tuxedo Park Association ran the community as a feudal estate—operating all utilities and assessing its own taxes—Loomis was in some sense lord of the manor. He had negotiated the shrewd real estate deal by which four thousand acres was annexed from the many warring heirs of the Lorillard estate and organized into the Tuxedo Park Association, with every family owning shares in accordance with the value of their property. He was one of the most powerful members of the association, at one point serving a term as president, and certainly knew how to use his influence to get his way.

As time went by, Loomis had less and less time for Tuxedo's stifling society. Even compared to the cashmere crowd in Newport and Saratoga, Tuxedo's tribal clannishness distinguished itself for its complacency and stilted formality. Even the doyenne of etiquette Emily Post, who grew up in the park and whose father was the architect Bruce Price, eventually tired of life behind a fence and departed. "Tuxedo people are not living from excitement to excitement," she once remarked sarcastically. "The fact that someone can and will give marvelous entertainments does not interest them in the least." Of course, Tuxedoites could afford to be unimpressed by most things, an attitude they evolved into a lifestyle. As Price Collier, one of the park's original residents, put it: "The best society of Europe is success enjoying an idle hour or so; the best society here is idleness enjoying its success. . . . Society, to be permanently interesting, must be made up of idle professionals, not professional idlers."

Tuxedo's virulent snobbishness did not appeal to Loomis, who never cared for fancy airs or anything artificial. "The beauty of it was that while he may have been born a WASP, he didn't think that way," said his grandson Tim Loomis. "Alfred was very curious about everything and everyone. He didn't have a belief system that bound him to anything. He was very much his own person at a young age."

For the scientists Loomis imported regularly, life at Tuxedo Park was as exotic and thoroughly peculiar as the prancing poodles and giant mastiffs they spotted roaming the grounds of the great estates. It was as if Loomis and Wood, who were both unabashed Anglophiles, had taken their lead from the great English savants and their turreted country estates and together created their fantasy of a private research laboratory with no luxury spared. Kistiakowsky fondly recalled the comforts of Tower House, which included an Italian chef, a Greek butler, and an English steward who claimed to be a Northumberland Percy:

Some years before I ever got there Alfred Loomis bought an older estate, with the main building the size of a large English manor house, located on the brow of the highest inhabited Tuxedo hill with the most spectacular view of Tuxedo Lake down below. This building became the Loomis Laboratory, owned, I believe, by a Loomis Foundation, the research rooms occupying the lower part of the building while the palatial common rooms and a dozen or so large guest rooms being in the upper stories. All of this was presided over by a major domo, a younger son of the Duke of Northumberland (it was a duke but which one I am not sure) who misbehaved himself in England so consistently as to have been dispatched on remittance to Canada; eventually he made his way to the Loomis Laboratory in Tuxedo Park. He treated us regularly to English country squire food, including English breakfasts and small fried filet steaks for lunch, the beef so well aged that they reeked to high heaven even after frying, but melted in the mouth. The duke's son was eventually fired for regularly embezzling the housekeeping moneys. . . .

The two senior regular guests were the great American experimental physicist Robert W. Wood . . . and a biologist from Princeton—Newton Harvey. The biologist was the great authority on fireflies and other luminous things and seldom talked about anything else than bioluminescence. Robert Wood was transformed from his professional self whenever pretty ladies were about, which was not frequent since the laboratory was an essentially male enclave, the wives being mostly only weekend guests.

The mainstay of the laboratory was Mr. Miller, a skilled mechanic but also jack-of-all-trades, who lived with his wife in one of the lesser buildings on the estate. . . . Stimulated by Wood, I got involved in spectroscopic work, using a very large (40 feet!) grating spectrograph built by Wood. Later, together with Alfred Loomis we designed a refined version of it, Mr. Miller and I built it and I photographed and measured the ultraviolet absorption spectrum of formaldehyde, as well as some spectra with which I had less success. . . .

We all worked industriously in the laboratory and then adjourned for a late afternoon high tea with English cookies and dainty sandwiches. After the tea some of us went back to the laboratory and then adjourned to one of our bedrooms to have some gin or beer. Later we had dinner in the dining hall, seated at several tables with four settings of real silver on each and finally had demi-tasse coffee back in the great hall. The mammoth reception room next to it was seldom used. The laboratory provided no liquor, it being the prohibition era. Expeditions down to the village for beer and gin at the bootlegger's were a standard distraction. Weekends, some of us went sailing, others playing tennis and golf (those I did not, but Bill did and was tops in tennis). Occasionally we all were invited to dinner at the Loomises where I got to know Ellen and their three sons. . . .

The village of Tuxedo Park provided few diversions for the young scientists. The narrow roads meandered through sixteen hundred acres of parkland, and they quickly tired of peeking through iron gates at the odd Italianate palazzo or French château. The hamlet itself resembled a quaint English village and consisted of little more than a train station, post office, and library. The latter featured the occasional speaker and had once even pressed into service Mark Twain, who had rented a house during the summer of 1906. According to the story told by Ellen Loomis, who served on the library's board of trustees, even this rare cultural offering failed to draw a crowd, and the author had gone home angry: "The Park people thought it was for the hamlet's benefit and the hamlet thought it was for the Park residents, so neither came." Ellen, like her husband, did not involve herself in Tuxedo's social swirl and had few close friends in the community. "Partying was not her recreation," observed one acquaintance, adding that because of her education and intellect, "many of the ladies in the Park were nervous in her presence."

Not surprisingly, the Loomises and their coterie of scientists were left pretty much to themselves. "The Tuxedo Park residents mostly regarded us as a leper colony or at best as Alfred Loomis' private toys," recalled Kistiakowsky. "Therefore, we spent much of the nonworking time in chitchat at the laboratory or reading."

Kistiakowsky's frequent visits to Tower House with Richards led to an invitation to spend the summer of 1930 as a research fellow at the Loomis Laboratory, and he would return for several more years. He later recalled that two entire summers were devoted just to the formaldehyde project Loomis had agreed to sponsor. The spectroscopic photography was "a lot of work," he wrote in his memoirs: "Each exposure lasted hours, during which time I had to watch that the formaldehyde vapor was at the chosen (constant) pressure and did not condense on the windows of the long absorption tube. And as often as not something went wrong with the spectrograph or the exposure was not properly chosen and so the photographic plate was useless." The measurement of the spectrum—"more than 100,000 individual measurements with the traveling microscope"—he did later at Harvard. He converted the readings to wavelengths or frequencies of the absorption lines, plotted these on rolls of graph paper, and looked for empirical regularities. These he then sent to Gerhard Dieke at Johns Hopkins, who had functioned as the theoretician on the project at Tower House and worked to fit the readings to the predictions of the still very youthful quantum mechanics. "We had considerable success," Kistiakowky noted, "for the first time the rotational spectrum of an asymmetric polyatomic molecule was used to calculate the interatomic distances and bond angles of the molecule."

During this same period, Loomis had Richards, who by then had participated in a number of different research projects, working on an apparatus that utilized short-wave radio to produce artificial fever and was used for medical treatment. The research came about as a result of an accident at the General Electric laboratory, where Loomis had many friends. Men working on a six-meter short-wave set had inexplicably fainted or become ill, and it was found that they were suffering from a high fever. The fever was traced to the effect of the short-length radio waves. Loomis was intrigued, and after further experiments at the Tower House laboratory, he and Richards discovered that the short waves had the power to increase the temperature of salt solution—but not pure water—and apparently acted on the salt in the bloodstream. Based on their experiments, which were later published in a paper, Loomis and Richards constructed an apparatus that would induce fever in carefully measured dosages to patients suffering from paresis, or partial or complete paralysis. The value of fever in curing certain infections had long been known, and it was not uncommon for patients to be exposed to malarial mosquitoes for such a purpose. The value of Loomis' "artificial fever," noted the *New York Tribune*, "is that it is more controllable and less harmful than the malarial fever, which had to be checked with quinine."

Loomis himself was a large, but by no means constant, presence at Tower House. He was at his Bonbright office in the city during the week and was often away for extended periods on business trips and holiday excursions. The day-to-day running of the laboratory was left to Garret A. Hobart III, the grandson of McKinley's vice president by the same name and a Tuxedo Park neighbor. Hobart, whom Richards describes in his novel as "a weedy young man of conventional good looks," was never particularly robust, and after graduating from the Choate School, he had become seriously ill. He enrolled at Johns Hopkins but was in and out of the hospital and completed only about a year of college. While he was there, however, he became an avid admirer of Wood's and stayed on as the physicist's research assistant for another year before being forced to leave because of his frail health. He spent most of his twenties apprenticed in the laboratories of General Electric in Schenectady.

Hobart was painfully shy and nervous but very bright, and he was passionate about electrical gadgets, an obsession he shared with his father. He passed most of his time building microscopes and tinkering with his short-wave radio sets in a workshop in his home. As he was heir to a sizable fortune, he had no need of regular employment, and on Wood's recommendation, he went to see Loomis in Tuxedo Park. Loomis took an immediate liking to him and made Hobart his protégé and secretary at the laboratory, where he proved extremely handy at all sorts of wiring and repairs. He and his lovely wife, a Belgian named Manette Seeldrayers Hobart, took a house only minutes from the Loomises in Tuxedo Park, and the young couple became a fixture at Tower House.

In the early days, the laboratory was very much a family affair. Many of the young researchers who came to stay earned their keep by acting as part-time tutors and companions to Loomis' energetic brood, who were becoming something of a handful for their mother. Ronald Christie, a medical student from McGill University in Montreal, spent several summers at Tuxedo Park and became virtually a member of the family. Christie was specializing in lung diseases, which appealed to Loomis, because it had been his grandfather's field, and he took the young doctor under his wing. The mornings were spent doing research in physiology—they published several papers together—and afternoons and weekends were taken up with swimming, sailing, and fishing for trout on the lake with Lee, Farney, and Henry and any other lab hands who cared to tag along.

With so many young men about, and so many idle country evenings, an inordinate amount of drinking went on. Prohibition had made alcohol a thriving cottage industry, and no self-respecting scientist could resist whipping up a bathtub batch of ethanol and juniper drops. Christie loved to tell the story of the time, shortly after first arriving in Tuxedo, that he became quite ill from the effects of the homemade hootch. Being a medical student, he was sure he recognized the symptoms of serious alcohol poisoning, including extreme dizziness and flickering vision. Thoroughly alarmed that he might have downed something lethal, he sought out Loomis and told him it was "not two hours after the cocktail hour that he had started seeing the flashes of light." After hearing him out, the much bemused Loomis informed Christie that the spots before his eyes were in fact lightning bugs and suggested the young Canadian might want to study the indigenous insect when he was sober.

Christie, like many of the young men who congregated at the Tower House, regarded Loomis as a mentor and second father. Loomis had an ease and warmth with them that he seemed to lack with his own boys, and over the years he attracted a devoted coterie of brilliant young scientists he came to call his "other sons." Almost fifty years later, when Loomis was old and beginning to fail, he would ask Christie, by then an eminent lung specialist at the Royal Victoria Hospital in Montreal, for a very personal favor. His loyalty and affection for Loomis undimmed by the years, Christie would oblige.

LOOMIS' business often took him to Europe, and in 1928, with his laboratory up and running smoothly, he felt free to take off on another scientific tour of the Continent with Wood. It was essentially a shopping expedition, and Loomis had an unlimited budget when it came to gadgets and experimental gear. At the top of his list was one of the famous astronomical "Shortt clocks," a new instrument for improving accuracy in the measurement of time invented by William Hamilton Shortt in 1921. Loomis' fascination with exact timekeeping was probably an outgrowth of his interest in navigation, a hobby he had cultivated since his boyhood days sailing the waters off Long Island. The Shortt clock had a "free pendulum" swinging in a vacuum in an enormous glass cylinder and was reportedly accurate to one-tenth of a second a year. It was so expensive that only the five biggest, best-endowed observatories in the world could afford to own one. But this only heightened Loomis' interest.

They made a beeline for London, where Wood took Loomis to the workshop of F. Hope-Jones, who made the clocks. When they climbed up the dusty staircase, they saw little in the way of machinery, but standing in one corner of the room was one of the superb clocks. It was almost completed, Wood recalled, "which made the total production to date six":

Loomis asked casually what the price of the clock was, and on being told that it was two hundred and forty pounds (about $1,200 at the time, which was roughly what the average American worker earned in a year), said casually, "That's very nice. I'll take three." Mr. Jones leaned forward, as if he had not heard, and said, "I beg your pardon?"

"I am ordering three," replied Loomis. "When can you have them finished? I'll write you a check in payment for the first clock now."

Hope-Jones, who up to then had worn the expression of a man "who thinks he is conversing with a maniac," was taken aback. He immediately became extremely apologetic and insisted that no payment was necessary until he had made good on delivery of all three clocks. But Loomis insisted on handing over the check, much to the other man's amazement.

While in England, they paid a visit to Sir Oliver Lodge, the eminent British physicist known for his pioneering research in radio frequency waves, who presented each of them with an autographed copy of his latest book, *Evidence of Immortality*. They then called on Sir Charles Vernon Boys, another noted physicist, who was famous for his highly sensitive instruments, including his invention of the radiomicrometer for measuring radiant heat, and an automatic recording calorimeter for testing manufactured gas. Loomis took an instant liking to Boys and invited him to return with them to the United States and spend the remainder of the summer in Tuxedo. Boys, who was then seventy-three years old, protested that he was pretty feeble to make such a journey. But Loomis urged him to accept and reassured him: "All you have to do is be in Plymouth on July 4, and I'll arrange everything else." They spent the last few weeks motoring around England, presented motion pictures of their experiments with super–sound waves before the Royal Society, went to the Derby, dined with "celebrities," and then flew to Copenhagen, where they met with Niels Bohr.

Their last stop was Berlin University, where they again showed the motion pictures of their experiments, and Wood introduced Loomis to "most of the other famous scientists then alive in Germany," including the botanist Nathanael Pringsheim, the physical chemist Walther Nernst, and two Nobel laureates in physics, Max von Laue and Max Planck. They also stopped by the University of Göttingen, and as Wood recalled, during the visit they were invited "to see a student duel." Wood was eager to accept the offer, but dueling was against the law, and Loomis politely, but firmly, declined. They began their voyage home on the *Paris*. Loomis had sent Boys his first-class steamer ticket, and sure enough, when the ocean liner slowed in Plymouth to allow the English passengers to board, Boys was waiting for them: "There he was waving his hand joyfully and all ready to scramble up the gangplank, looking as relieved at finding us really on the steamer as were we at seeing him on the tender."

Loomis saw to it that they had "the best of everything on the boat."

The chief steward prepared a fine French dinner for them every night. On the last day, he announced that he had prepared a grand surprise for dinner, something that was very unusual, he promised, "a great luxury!" That night, after the soup and fish courses were cleared, he solemnly rolled a wagon over to the table, bearing a large covered silver dish. It was a meal Wood never forgot: "He rubbed his hands together and smiled at us, and then lifted the cover, displaying in all its stark nakedness a huge shapeless mass of shivering, steaming corned beef, garnished with cabbage and cauliflower and whatever else goes with this, my pet abomination, a New England boiled dinner."

Upon their return to Tuxedo Park, Loomis and Boys spent the rest of the summer doing lightning experiments, using a special high-speed camera designed by Boys expressly for photographing rapidly moving objects such as bullets and lightning bolts. Together they succeeded in taking a series of photographs that proved Boys' theory that the path of the lightning bolt was cleared by an electric "beam" that preceded the bolt. Although they did not have a hand in creating the dark thunderclouds that hung low over Tuxedo's rolling hills, many of the residents reportedly blamed the atmospheric disturbances on the activities up at Tower House.

When the Shortt clocks finally arrived, Loomis installed them in a vault excavated from the solid rock of the mountain on which the laboratory stood, mounting them on three massive masonry piers that were, in effect, part of the bedrock. The location was "especially favorable," according to Loomis, because it was practically free from traffic and electrical disturbances and was carefully temperature controlled. This meant the fourteen-pound pendulums were swinging in a near vacuum and in planes 120 degrees apart. Loomis then went to Bell Laboratories and bought "the best" quartz crystal clocks they made to use for comparison purposes. These one-hundred-thousand-cycle quartz oscillators, invented in 1928, were accurate to one second in thirty years and were built primarily for the U.S. Bureau of Standards. The advantage of the quartz crystal clocks was that they were inherently stable and relatively free from extraneous effects. Because they were not dependent on gravity, they could, without any adjustment, operate at the same rate in any latitude and at any altitude.

Equipped with the most accurate, reliable—and expensive—clocks then available, Loomis began collecting all sorts of data it had been previously impossible to record. He was able to achieve spectacular results and to publish major findings, in part because quartz crystal clocks were still such a novelty and few scientists at the time had access to the superb assembly at Tower House. Loomis proved that there was no such thing as keeping perfect time, showing that even the five most accurate clocks in the world—the three in his vault and the other two in the Naval Observatory and in Greenwich, England—were subject to numerous errors. All that was possible was to make comparisons between the different clocks.

Loomis set up such an accurate system of precision clocks and comparative time recording that he was able to register infinitesimal fluctuations in the clock rates due to the fact that the pendulums coupled and influenced one another despite all the precautions taken against disturbances. Describing his test at the winter convention of the American Institute of Electrical Engineers in January 1932, Loomis said, "This would seem to show that, massive as the piers have been made, they are not infinite in comparison to the fourteen-pound pendulums, and that strains are set up by each pendulum that are felt in some degree by the others through the piers and solid bedrock." He found that when he placed the clocks at the corners of an equilateral triangle, facing inward, the coupling was broken. He even found a way to calculate the change in gravity pull if a clock was raised one foot and concluded that the loss as a timekeeper would be one and a half seconds in a year.

Over the next few years, Loomis and W. A. Marrison, a researcher at Bell Labs, conducted a series of important experiments comparing the performance of Loomis' free-pendulum clocks and the quartz clocks at Bell Labs in New York, some fifty miles away. Loomis installed a private line that carried the Bell oscillator signals, and he designed an ingenious chronograph to compare the timekeeping abilities of the Shortt pendulum clocks and the quartz oscillator clocks. The comparison was effected through the circuit maintained between the two laboratories, over which a one-thousand-cycle current, controlled by a crystal in New York, was used to drive the Loomis chronograph. Because the pendulum clocks were sensitive to the pull of gravity but electric clocks were not, Loomis used his chronograph to show the moon's effect on pendulum clocks—something that was known but had never actually been demonstrated before. Boys told the *New York American* that the Loomis clock had "shown that compared with it the ordinary pendulum clock, of the finest sort, goes wrong daily on account of the gravitational pull of the moon in six hours."

The experiment required that he simultaneously record for several weeks on miles of tape the minute variations in the time shown by the gravity clocks in the different locations. As this was long before the advent of computers, to help with the painstaking data analysis of the tapes Loomis hired a battery of women who operated desktop computing machines and ran the numbers. The figures were then studied by Ernest Brown, the eminent Yale astronomer, who confirmed the distortive effect of the moon's gravitational pull on earthbound gravity clocks. Loomis published the final results later that year in his paper "The Precise Measurement of Time," in the *Monthly Notices of the Royal Astronomical Society*, followed by a paper with Brown's findings.

Brown, in remarks before the winter convention of the American Institute of Electrical Engineers, explained that time "is relative in more than the purely Einsteinian sense." Accurate time could be obtained only by comparing our clocks with a standard clock, but the standard itself was subject to various errors. Some of the sources of error were known and could be adjusted for, but there were many other causes—terrestrial and celestial—that act as "time thieves." Loomis, Brown reported, had "just lately" caught the moon stealing time from the earth: "For the first time the action of the moon, which is the greatest external effect, was measured by the Loomis chronograph and shown to give accumulated errors which were always less than two ten-thousandths of a second as indicated by theory."

Loomis' work won him memberships in both the Royal and American Astrological Societies. He became so obsessed with precise time, he kept a radio set in the basement laboratory that automatically tuned in to Berlin, Greenwich, Arlington, and the other observatories just in time to catch their time signals, which were then recorded. Loomis would read the record of official time signals, compare it to his own clocks, and, according to one observer, "predict with a chuckle the exact corrections which the various observatories would have to broadcast at the end of the month." One New York paper ran a brief story that gently spoofed the latest obsession of the "eminent American capitalist whose one passion in life is to conduct and promote subtle scientific researches":

What time is it?—you ask.
It is p.m.—hour 12, minutes 2, second 0.0003.

Loomis would remain a "time nut" for the rest of his life, according to Luis Alvarez, who recalled that Loomis always wore "two Accutrons—one on his right wrist and one on his left wrist." He would check them every day against WWV (the standard frequency broadcasting station of the National Bureau of Standards), and if one was gaining a half second on the other, he would wear it on the outside of his wrist instead of the inside, so that gravity changed the rate of the tuning fork and the two watches tracked each other, and WWV, "to within less than a second a day."

Loomis' scientific investigations followed a pattern in which he set out in one direction after another in search of a new discoveries, seemingly only to abandon it. Even to the most casual observer, his feverish efforts, followed by a brief triumph and equally feverish desire to be off again on another tangent, must have appeared somewhat self-indulgent, even frivolous. There was also a constant shuttling of European scientists and experts back and forth across the pond, for he needed playmates. After a while, it seemed that Loomis' attention was as transient as his guests. Who could have known then that it was fortunate that he would give his imagination such free rein—from his earliest explorations of high-frequency sound waves to his chronograph and experiments with quartz crystal clocks—for it would lead him into his research of the nascent field of radar, which would become critical in the coming war.

The day after Loomis' forty-first birthday, Ellen wrote to Henry Stimson that she sincerely doubted her husband, "who is getting all grey over the ears, though it is becoming," would ever stop playing with his mechanical toys and grow up:

> Not that it makes any difference, he will always be a boy, no matter how brilliant he is, no matter if he lives to be 98, as Stimsons should. Henry [their youngest son] consulted as to when we could celebrate his birthday with appropriate ceremonies of cake, and candles and gifts. Alfred is so surrounded and encompassed with scientists all the time nowadays. Lunch, tea, and dinner yesterday were full of short wave radio men, marine bacteriology men, and chemists, that we held our little private festival at breakfast. . . .

In another letter to Stimson after Christmas, Ellen's loneliness was apparent in her touching description of having her boys back at home for the holidays. " 'Heaven on Earth!' is the only way I can describe them!" she wrote. "With riding, skating, and long hours with their father in the laboratory, and much reading aloud with Father, Mother and me, the days were all too short. And now those blessed weeks are all over." She had renewed her efforts to participate in her husband's life. Although in the past she had complained of being "a little hazy about almost every subject Alfred talks on, either financially or scientifically," she now recommended a thick book on astronomy she was reading, "written so that a plain person could more or less understand it." She confided: "Of course, it interests me as a light on the world of thought Alfred lives in." Loomis, she wrote, was preoccupied with his research:

> The Tower was full of scientists and Alfred has been keener than ever on the problem of actually "exact" time. He has had all sorts of adventures among his clocks, and learned a lot, and there is plenty left to learn for a lifetime ahead. Also, he is making a new and improved model of his "interferometer" for measuring the resistance of fluids. He has been elected to the council of the American Physical Society—it is a real honour and he is pleased. . . .

Working in his laboratory alongside accomplished scientists, Loomis excelled as the innovative designer of precision mechanical devices. He did not have the patience to do the involved analyses accomplished by some of the brilliant researchers who came to Tower House and instead made his contribution to these studies, as Kistiakowsky put it, by "building complex apparatus" that advanced the basic knowledge in the field. What followed was a string of inventions of the kind he had loved coming up with since boyhood. Working with John C. Hubbard, a visiting professor from Johns Hopkins, Loomis developed a "sonic interferometer" to analyze the molecular effects of supersonic waves in liquids.

Another Loomis gadget, and one of which he was especially proud, was "the microscope centrifuge," developed with E. Newton Harvey, who was a professor of biology at Princeton. The device enabled biologists to witness for the first time what happened to cells when they were subjected to high gravitational forces and led to new discoveries in cell structure. As Loomis and Harvey wrote in the introduction to their first paper on the subject:

The previous procedure has been to centrifuge the cell in a capillary tube, remove it from the tube and observe it under a microscope to determine what happens. It would obviously be far better to observe the effect of centrifugal force while the force was acting. An instrument for his purpose could be constructed in theory, making use of several different principles. Our communication describes a practical means of attaining this end.

To create the microscope centrifuge, Loomis and Harvey adapted the principle of the motion picture projector. While the cell was being whirled around at a rate of eight thousand to ten thousand revolutions a minute, their instrument presented a series of images with such regularity and rapidity that it appeared to the observer as a clear and steady picture. In the microscope centrifuge, a disk, similar in size and operation to a turntable, was rotated at high speed by an electric motor. They placed the slide containing the cell or egg to be studied on the disk and focused the microscope on a given point on the slide. Each time the disk completed its revolution, the slide came into the field of vision, and at that precise moment a light lasting one one-millionth of a second flashed. The flash was produced by a small mercury light, the frequency and duration of which were controlled by the discharge of electricity through the mercury vapor. The rest of the revolution occurred in darkness, thus creating the illusion of a continuous picture. Loomis, the magician, was at work again. "So with this new microscope," concluded *The New York Times*, "we have a remarkably ingenious application of a familiar principle to a new purpose."

Loomis and Harvey produced a series of photographs that showed that when rotating at such a high speed, the cell or egg was subjected to terrific centrifugal pressure, what is now referred to as "g-forces." As a result, the granules and the internal structure of the cell underwent various distortions, and the egg changed shape and finally broke up. From observations of the tension at the surface of the egg and behavior of its internal parts, it was now possible to ascertain facts about the fundamental characteristics of cells that could not be previously obtained and were of importance in biological research. Until their new microscope, for which they received a patent, scientists had been handicapped by their inability to observe and measure the various steps in the deformation of cells and in the movement of particles within them. Almost immediately after the publication of their first paper, some existing theories about the properties of matter within cells had to be revised. Loomis and Harvey were later awarded the prestigious Wetherill Medal of the Franklin Institute for the microscope centrifuge.

As was his habit by now, Loomis put his name only on the first of the thirteen microscope centrifuge papers published by his laboratory: this was partly from modesty and partly because he had already moved on to the next project. The *Herald Tribune* applauded the latest in a series of Loomis inventions: "Further knowledge concerning the fundamental characteristics of cells, such as the eggs of marine life, has been obtained, it was announced here, by a new type of microscope. . . . Mr. Loomis, the banker-scientist and collaborator of Dr. Harvey in the study, has developed an important laboratory at Tuxedo Park, which has been of value in throwing light on several important scientific problems in the last few years."

Loomis was devoting every spare moment to his research at Tower House, and the toll it was taking on his energy and attention did not escape Stimson's notice. In the spring of 1928, after learning that Loomis had succumbed to a bout of influenza and had spent several weeks resting and recuperating in Florida, Stimson, who was then abroad serving as governor general of the Philippines, sent him an admonishing letter:

> My main anxiety with regard to you is the fear that you burn the candle at both ends, and I beg you not to do it. I am glad to hear that everything is going on well with Bonbright, and I hope that it continues to do so. I see there has been, and still is, a big boom in the stock market which, so far as I can see, is not supported by the general condition of business. I trust you are all keeping your watchful eyes upon that situation. . . .

Stimson's concern was not solely for Loomis' welfare. He had long ago entrusted his favorite cousin with managing his finances and feared Loomis' many distractions might not bode well for his own interests. Bonbright's long run of good fortune had been Stimson's as well, making him by his own estimation a rich man and underwriting a lifestyle his years of public service would never have afforded. Part of his purpose in writing Loomis was to ask for his help and advice in courting Wall Street. The Philippines was in desperate need of economic development, and Stimson, a devoted corporate capitalist, believed large quantities of American investment was the answer: "By bringing in public utility companies that offer their services first and depend upon making their own market by the demand which they create, I hope to turn the flank of the deadlock and get the Filipinos to understand the real benefits of modern American capital and methods. . . ."

Stimson need not have worried about Loomis' priorities. While his passion was science, he was far too competitive and too committed to his company's success to allow himself to be distracted for long. He remained deeply involved in Bonbright's day-to-day transactions, his brilliant analytical skills proving the perfect complement to Thorne's keen judgment. When the direction of a merger needed leadership and vision, Bonbright was seen as always at the ready, having established a reputation for undertaking swift negotiations and acting with consummate entrepreneurship. In addition, Loomis sat on the board of half a dozen companies, and most of what there was to know about what was happening was his to know. As Loomis reassured Stimson in August 1928:

> Things at the office are going along according to schedule.
> Superpower is in a flourishing condition and its stock is now selling at 44, after the dividend of $20 per share of Preference stock, which amounts to 64 for the old stock.
> The new company that we formed, Allied Power & Light, as a merger . . . has worked out extremely well and both organizations have gotten to like each other more and more, and I think the merger will turn out to be a very constructive step for all of the public utilities under the influence of these two groups. . . .

He mentioned Tower House only as a cheerful aside: "There are eight or nine scientists staying at the Laboratory and the work there is as interesting as ever." While Stimson continued to beseech him for news of his work with Thorne at Bonbright, Loomis' letters are sparse and hurried: "I am just rushing off to a meeting of American Superpower so I won't have time to say anything more in this letter," he explained in a brief note on September 5, 1928. "Things are going awfully well with that company and all our plans are moving along just the way we wish."

By early 1929, Loomis and Thorne had pulled off what was in many regards its largest and most important enterprise, merging three important holding companies into a new group known as the Commonwealth & Southern Corp. On January 11, Stimson received from Ellen a glowing report of Bonbright's success: "Alfred has been having thrilling days," she wrote, referring to the prospects for another new utility holding company they were putting together, the United Corporation. "Alfred and Landon have had a most interesting time, and enjoyed every minute of it."

While the two brothers-in-law divided the team work according to their individual talents, they were of one mind when it came to doing everything together on a partnership basis. This included adhering to a strict policy of keeping the bulk of their profits in cash. Bonbright, unlike most other investment houses, never carried large inventories of the securities it underwrote—which would prove to be the undoing of many of the biggest promoters of the bull market. With the stock market shattering record after record, paper speculation was spiraling higher and higher, and most of their competitors were betting confidently that the country's new prosperity would never end. The boundless optimism reached to the highest levels of the government. On October 24, 1929, Thomas Lamont, one of Herbert Hoover's chief advisers on Wall Street, sent a memo to the White House in which he dismissed the skittish president's fears about the wild marketplace and, citing the success of new holding companies such as the United Corporation, reassured him that "the future appears brilliant."

But Loomis and Thorne were chary of the speculation fever and, like Stimson, doubted it could continue on unchecked. After all, they had been the architects of the so-called second industrial revolution of the twenties: they had helped build the generators that powered the big new automobile, telephone, and appliance factories that fueled the thriving consumer economy. Experts in the law of supply and demand, they understood that the conveyor belt factories that had proliferated across the country were now in danger of saturating the market and watched with foreboding as first automobile sales slumped and then department store earnings dipped for the first time. Loomis would later maintain that everybody on the Street knew the crash was coming, the only difference was that he and Thorne refused to bank on its being inevitably delayed.

When the house of Morgan, which considered itself the king of the Street in those days, decided to enter the public utility field for the first time, it did the unheard of—and asked Bonbright to act as its ally and agent. Because of their broad experience in power mergers, Loomis and Thorne were able to bring together the different entities and incorporate United as the Morgan-Drexel-Bonbright holding company in record time, without the usual rancor and difficult, drawn-out negotiations. The United Corporation literally united under one umbrella the giant Mohawk-Hudson Corp. of New Jersey, Columbia Gas and Electric, and a string of smaller utility companies—in all, controlling more than a third of the power production in twelve eastern states. Its board was filled by Loomis' and Thorne's friends and Morgan men and was as incestuous as could be.

At the time of the offering in January 1929, Loomis and Thorne suggested a value for the stock that they thought the market would bear and were surprised and somewhat uneasy when Morgan, which had been caught up in the flurry of stock promotion, insisted on a higher price. But Loomis and Thorne masterfully orchestrated the financing, and so persuasive were their arguments in favor of expanding ownership and unifying control, they easily effected the offering at the higher price. The initial market value of the new United Corp. was $260 million. "It was viewed as a great triumph at the time," said Ed Thorne, "but my father and my uncle realized that the price was unreal—the value wasn't there behind the stock. After that, they came to the conclusion that the market was completely out of hand and they said, 'Let's sell out.' "

Very quietly, over a period of months, Loomis and Thorne liquidated their remaining securities and converted everything into sound, long-term treasury bonds and cash. Then came Black Thursday. When the market broke on October 24, 1929, stock prices plummeted, and stunned investors, many of whom had been buying shares with borrowed money, saw their entire net worth wiped out overnight. As panic seized Wall Street, not everyone looked kindly on the two prudent financiers who were conveniently caught with their "pockets full of money," as Ed Thorne put it. Others in their class did not fare as well: the Vanderbilts lost an estimated $40 million of their railroad holdings alone, and J. P. Morgan Jr. was thought to have lost at least that much, if not more. It did not help that in the midst of so much despair, with the economic situation deteriorating day after day, Loomis and Thorne continued to profit handsomely. From that time onward, Loomis' critics would always try to cast a shadow on his prescience, implying that his good fortune in such a dark time could only have been the result of some extraordinary deviousness on his part.

Maybe they were "just lucky," as Landon Thorne always maintained, or perhaps, as Loomis would later claim, the mathematical charts he devised to follow the market did not justify gambling on a bell curve. The fact that Loomis made an estimated $50 million during the first few years of the Depression served only to intensify the mystique about the "scientific approach" he used to guide his financial affairs. He was among those few Street veterans who had foiled Black Thursday— Bernard Baruch always claimed that he saw the Crash coming and made a bundle shorting stocks he later bought back at a profit—and his foresight was now legendary. According to one popular theory recounted in *Fortune* magazine, and which may or may not be apocryphal, Loomis adapted the standard biological growth chart to the problem of timing—namely, when to get into and out of businesses:

> Loomis, it seems, once sat down with graph paper, pencil and documentary material to plot the "growth curves" of various industries along a time axis. According to the story, he decided from this comparative analysis that the time to put your money into any particular industry is while its growth curve is still on the upgrade, and that the time to cash in is just before the curve starts to level off. On this theory, he is also supposed to have decided that public utility investments were the thing to get into in 1920 and to get out from under in early 1929. Some market operators might consider this just a fancy way of saying that it is important to buy at the low and sell at the high, might even raise the point that Mr. Loomis had one of the best inside tracks in Wall Street at the time. . . .

Loomis and Thorne went into the Depression period with all their balances in cash. Despite "some arguments between them about whether they should get back into the market," they held their position, insulating themselves from the brutal battering investors took again in 1933 and 1934. They also shrewdly advised those closest to them, including Stimson, who was then serving his second stint as secretary of state, this time in President Herbert Hoover's beleaguered administration. Stimson, who traveled regularly to New York to consult them about the country's economic problems, as well as his personal finances, found them "well-heeled and prepared for the future":

> Alfred and Landon had been very wise in the way they handled affairs over the depression, and incidentally handled mine, and the situa-

tion is not discouraging, although nobody knows what is going to happen. They have their own formula and it is a wise one. Having in mind the possibility of inflation and the effect that would have upon long-time investments, they have, therefore, kept their investments all in a very liquid condition, like a boxer in the ring ready to meet danger from any direction. . . .

Even in a crisis, Stimson found he could always count on Loomis; his loyalty and sound judgment were unshakable. "I always have a feeling of the affection which Alfred Loomis has for me," he noted in his diary. "It is a very comforting thing." To his surprise, Stimson discovered he liked Loomis' smooth partner better than he had first judged. Following "a very satisfactory talk over business conditions" with Thorne, who had obligingly driven out to Highhold from Bay Shore early on a Sunday morning to meet with the busy secretary of state, Stimson conceded, "He is a very thoughtful, high-minded fellow, and I was glad to find that his views, hopes and ideals corresponded very closely with mine."

If Loomis emerged from the Crash in an even stronger position, the opposite was true of the majority of Wall Street bankers and stockbrokers who populated Tuxedo. The Crash wiped out huge fortunes and devastated the community that had once been a smug citadel of inherited wealth. As the Depression took hold in the early 1930s, many families departed from the park, and some of the great mansions were boarded up to reduce maintenance costs or even burned down to save on real estate taxes. Loomis reportedly bailed out several neighbors and made loans to others to help tide them over. If life was difficult for Tuxedo's elite, it was bleak for the villagers who toiled inside the park, who had for generations been dependent on the residents of the big manor houses for their livelihood. Loomis was more sympathetic to the workingman's plight than some of his distressed and penny-pinching peers. In 1930, at a meeting of the Tuxedo Park Association, the general manager suggested that wages for employees be reduced from $3.50 per day to $3.00 as a way to economize. When Loomis asked him if he thought he could live on such a wage, he replied that he could not, but then, "he was different." Loomis objected that he could see no difference, and the wages were not cut.

Loomis' own household was not immune to the effects of the Crash. According to one story, a number of the cooks and maids who had served at the elaborate parties at his home and at Tower House had made a practice of listening to the dinner conversation for stock tips and had acted on what they heard—several of them forfeiting their life savings. Gambling on the markets had been the contagion of the Jazz Age, and the women in his private employ, it seemed, were no more immune than the men on Wall Street. When Loomis learned that several of them had lost substantial sums, he immediately called in all seven of his staff, gave them each $1,000, and sternly lectured them "never to invest on the basis of gossip."

Chapter 5

CASH ON THE BARREL

Once he's made his mind up about something there's no talking to him.

—WR, from *Brain Waves and Death*

I N the years following the Crash of 1929, Loomis and Thorne turned increasingly to the field of investment banking, becoming the dominant players in the Bankers Trust Company. They were also the major powers in the Central Hanover Trust Co. and in the First National Bank, which was the old institution of George F. Baker, the "Titan of Tuxedo," as Cleveland Amory dubbed him, and one of the richest men in the country until his death in 1931. Thorne, who was close to the Baker family, was appointed a director of both Bankers and First National and joined the board of First National. Loomis became a trustee of the Central Hanover, as well as a member of the executive committee of Bankers Trust. While the Bankers Trust, First National, and Central Hanover were not the biggest banks in the city, they were among its most influential. By the beginning of 1932, with their many directorships and overlapping spheres of influence, Loomis and Thorne were towering figures on Wall Street. The *New York Evening Journal* reported: "[They] are one of the two or three most important financial interests in the banking business. The others are the Rockefellers, with their interest in the Chase National, and the Morgans."

As the economic consequences of the Depression necessitated, the two financiers took an active role in trying to restore confidence in the banking industry and avoid a collapse. After a dinner with President Hoover at the White House on January 4, 1932, Stimson raised the matter of "a new proposal for a bank bill" that had come to him from Loomis:

> Alfred Loomis, Landon and George Roberts [senior partner of Winthrop, Stimson, Putnam & Roberts] are all very much troubled over the banking situation and are very much in fear of a really big collapse of all the banks, and they have been working over with the officers of Bankers Trust Company a plan, rather ingenious in its nature which I think Alfred has been at work on, which amounts to an amendment of the banking law and will permit the officers of a closed bank to use its liquid assets to form a new subsidiary and go right on in business, making the new assets available at once for distribution among the depositors and thus restoring credit. The President was at work on the same thing from another aspect. That evening when I was with him, he was talking over the telephone with Barney Baruch and others. So that this suggestion fell right in with his thoughts. He authorized me to arrange an interview between Alfred and George Roberts and Ogden Mills of the Treasury. . . .

When he got back to Woodley, his baronial estate on the outskirts of the city, Stimson phoned Loomis and company to tell them to "come right on to Washington" to meet with Mills. The evening papers carried a report of the emergency message Hoover had sent to Congress laying out all his defense measures against the panic. Loomis' proposal, Stimson confided in his diary, "came right in time." He added: "Everybody is working under pressure now, so I was glad to have my chance to help a little bit."

It was a bleak winter, and as the banking crisis became acute, Loomis found himself frequently called to Washington by Stimson, Mills, and other Republican advisers who anxiously sought his advice. Stimson continued to lobby vigorously for Loomis' amendment to the banking law. He also consulted Loomis, who had close contacts in British banks, about the worsening situation in Europe. There were dreadful reports from Germany, which showed ominous signs of being economically depleted and at the end of its rope. Britain, which regarded Germany as an economic bellwether, was contending with the loss of confidence that had spread to its own banking system, where there had been a run on the pound and the abandonment of the gold standard. Both Stimson and Loomis understood better than most in Hoover's administration that the fate of American finance was intricately tied to Europe's. If Germany fell into bankruptcy, it would be vulnerable to the forces of revolution, either on the Right or the Left, and that could only have grave implications for the rest of the world's fortunes.

Europe's travails, however, were secondary to the desperate situation at home. Ten million people were unemployed, and thousands of banks had gone under, with more failures threatened every day. Hoover, who ran for reelection in a climate of fear and confusion, lost by a landslide to Franklin Delano Roosevelt. Distracted by the demands of a brutal campaign, Hoover had overlooked the pressing problem of the war debts, and the half yearly deadline for payment was set for that December 15. During the summer of 1932, the European nations had negotiated a revision of the German war reparations and reached a "gentlemen's agreement" that this revision depended on the United States' willingness to accept changes in the European debt structure. Unfortunately, by November 1932, Congress was in no mood to review the debts and defer payment. Two days after the election, both France and Britain delivered notes indicating they might default on their debt payments. The news hit "like a bombshell," and Stimson would spend his final days in office trying frantically to negotiate a workable compromise. The American banks that had extended the credit reacted in panic and demanded help from the White House. During the tense four-month interregnum, Loomis commuted back and forth to Washington, trying to help Hoover's lame duck administration push through last minute legislation to rescue the banking industry.

On February 14, 1933, disaster threatened when Michigan closed all its state banks. Stimson wrote in his diary the following day that he feared there might not be enough time to implement Loomis' measures:

Alfred Loomis lunched with us after a long talk at the Treasury all this morning. He had been in a conference with Ogden Mills, [Arthur] Ballantine, and [Eugene] Meyer of the Federal Reserve Board, and others. Then he went back there afterwards and was with them all afternoon and spent the night again with us. The situation is very gloomy. They have decided they cannot get through the measures that Alfred and Landon had proposed to permit failing banks to segregate their good assets and go on in business with those, because it would have all kinds of bad amendments stuck on it in the House of Congress. So they were trying to work it out a different way, and by nightfall, they had devised a plan by which they thought it could be done through the medium of legislation in the different states aided by a joint resolution from the Federal Government permitting the Comptroller to do things in every state which the state law allowed. . . .

By the end of the week, Stimson's worst fears were confirmed. None of the requisite steps had been taken, and when the banks opened again, there would still be no relief in sight:

The talk at Cabinet Meeting this morning was about the situation at Detroit, which has become very serious and has not yet been settled. . . . Everybody now agrees that the proposed plan of legislation I brought forward a year ago from Alfred Loomis and Landon Thorne is the solution of the general situation throughout the country, but Mills reported that [Democratic senator] Carter Glass would not agree to it in the Senate and that blocked the whole situation. . . .

It would be left to the new administration to take action. On March 3, the night before Roosevelt's inauguration, Stimson called Loomis, who was back in New York, and they had a long talk. Loomis was "a good deal worried" and said he thought the approaching weekend would be "the critical one, and we will know better after Roosevelt's announcement tomorrow what is going to happen and how we are going to weather the storm." Loomis felt the great danger was that there could be a bad inflation that would upset values and destroy national credit. "That is what everybody is afraid of," observed Stimson, "and nobody has been able to get Roosevelt to take a strong position on it."

The day after he took the oath of office, Roosevelt called Congress into an emergency session in four days' time, suspended gold convertibility and gold exports, and declared a four-day banking holiday to stop a run on the banks. On March 9, Congress passed the Emergency Banking Act of 1933, which let the relatively healthy financial institutions reopen their doors, boosted the public's confidence in the banking system, and bought Washington the time it needed to enact legislation. Over the next one hundred days, from March to the middle of June, Congress would pass a raft of bills that would make sweeping changes in the banking industry, most notably the Glass-Steagall Act, which stabilized the country's banks by guaranteeing deposit insurance. Many of Senator Glass' measures took direct aim at the Wall Street giants and, to many in the industry, seemed designed as much with an eye toward punishment as toward reform. Loomis had been right when he had told Stimson that that weekend would be "the critical one." It proved to be the turning point in the country's fortunes. Roosevelt appealed for calm in the first of his famous "fireside chats" over the radio, and when the reorganized banks opened the following Monday, the lines of panicked investors were gone.

Despite the fact that they were from opposing political parties, Roosevelt asked the outgoing Republican secretary of state to brief him, and the two met and talked on several occasions. On March 28, Stimson went to see the new Democratic president at the executive office. Winthrop Aldrich, president of the Chase Bank, was present and engaged in a talk about the new restrictions on commercial banks and trust companies and how they should probably apply to private bankers as well. After Aldrich left, Stimson congratulated Roosevelt on his success in stopping the financial panic thus far and warned him against the damage that might be done if the further necessary legislation was taken up "by the leadership of the senatorial investigation rather than the direct executive leadership of himself." It is impossible not to detect Stimson's concern that the public was in a rush to blame bankers and exact revenge for Black Thursday. High finance was on trial:

> I said that in the ticklish situation of public confidence today such an investigation might do serious damage to public confidence as well as injustice to the bankers who had already put their houses in order, because it would make no discrimination between the innocent and the guilty. Roosevelt very warmly agreed in this and said it must always be borne in mind that most of the bankers were conscientious and were doing their duties. . . . We took up the subject of foreign affairs. I praised his banking bill, pointing out the importance of the final paragraph suggested by Loomis and congratulating Roosevelt on the

fact that the bill had stopped withdrawals and had produced an immediate return of gold and hoarded currency. . . .

Loomis, although politically conservative, was able to look past party loyalties to the candidate and his positions. He was sympathetic to some of the economic measures Roosevelt was trying to implement, particularly fixing the gold standard, which he told Stimson he "strongly believed in." Loomis participated in a series of "serious talks" among Stimson, Ogden Mills, economic adviser Herbert Feis, and Senator F. W. Walcott regarding Roosevelt's inflation bill, to which the Republicans were organizing an opposing statement. Loomis had been drawn into the matter after he received a telegram from Senator Francis Townsend asking for his views on inflation. After consulting Stimson, Loomis prepared a memorandum on his views of the gold bullion standard, in which he argued that "the old Peel view of currency standard," based upon a fixed amount of gold in the dollar or pound, was now obsolete in England. He sent the memorandum to Stimson, along with a hasty note:

Landon and I had a most interesting talk at lunch with Mr. Fred Kent, who, as you know, is the Foreign Exchange authority through which the Secretary of the Treasury is acting in controlling all foreign exchange operations of the country. He feels the same way we do about the proposed emergency gold bill, namely, that it probably offers the best method of solving our international situation at the present time.

Over the telephone, Loomis told Stimson he thought that "the legislation was pretty good in its final form." After reading Loomis' memorandum, Stimson passed it around:

I showed his memorandum to Feis, who said he agreed with every line in it except that he [Feis] might have gone a little further than Loomis in warning against the misuse of the powers granted in the bill. I told Mills over the telephone that Loomis had expressed approval to a certain extent of legislation and that I was a little sorry that Mills had not seen him or learned of his views before he advised the Republican statement in opposition to it. Mills was much surprised at Loomis' position. The issue in general seems to be that

Loomis and Feis and Roosevelt's advisors are in favor of a controlled
money standard. . . . I talked over Loomis' views with Walcott, and
Walcott said he agreed with them all except he did not approve of the
provision in the bill which gave the President Power to fix the gold
content of the dollar. . . .

On April 9, exactly one month after the Emergency Banking Act
was passed, Loomis and Thorne announced that they would be stepping
down as chief executives of Bonbright. Their "retirement" had essentially been forced by the bill, which prohibited investment bankers
from being directors of Federal Reserve Member Banks. By leaving Bonbright, they would be free to hold their many bank directorships without conflict with the new stricter provisions. Loomis and Thorne's
abrupt departure made headlines and was hailed as a seismic shift in the
structure of the financial world. In the peerage of American businessmen, they were among the sharpest players, distinguished for building
companies of rock-bottom strength and rapidly increasing capital girth.
They were the young and restless forces behind the giant new power
companies, and their impeccable credentials had allowed them to pool
money from a dozen of the greatest private fortunes in the country—
Morgan money, Mellon money—with the money invested by millions
of working-class people. Bonbright, thanks to their efforts, was now one
of the "Big Six" Wall Street investment houses, and their handpicked
successors, Sidney A. Mitchel and Pearson Winslow, could be counted
on as loyal allies in any future ventures. Most of their peers predicted
Loomis and Thorne would give up the securities business to move ag-
gressively into commercial banking, and as the *New York Journal* reported, there was considerable need for men of their proven ability:

> [They] are two of the small group of younger leaders who have come
> to the front in Wall Street in the last two years. It was in the 1920 de-
> pression that they took hold on Bonbright & Co. and steered it from
> the shoals of bankruptcy, toward which it was heading. . . . They did
> and, in effecting a reorganization, among those to retire were Senator
> Walcott of Connecticut, who started the recent stock market probe.
> Now another depression has rolled around and again Thorne and
> Loomis find themselves the gainers. . . . Both have done exception-
> ally well in this depression.

But that spring, Loomis informed Thorne that he was through. There was a certain finality about this announcement so that Thorne knew it was pointless to object. "They really did everything together, so when Alfred decided he wanted out, that was it," said his daughter-in-law, Betty Loomis Evans. "Alfred just totally lost interest in business. He felt he had enough money to do whatever he wanted, and what he wanted to do was science. In the end, I think Landon understood. They never had any words over it, or any kind of falling-out. Landon knew that was just who Alfred was."

Loomis was not someone you could argue with. He would listen patiently to an opposing opinion, reflexively patting his shirt pocket where he kept his Lucky Strikes and then lighting one up. But his consideration was nothing more than that—an act of politeness on his part. "He would always just shrug and walk away," said Evans. "I never once knew him to change his mind about anything."

According to Henry's first wife, Paulie Loomis, "Alfred was always set on being a scientist, right from the beginning. But he had responsibilities—he had to take care of his mother, his sister, his family. He helped Henry Stimson, and made him a lot of money. Stimson could never have had the career in politics he did without it. Alfred was a very premeditated person. He had it all figured out. He did what he had to do, and the first real chance he had, he took all his money and invested it in his own firm and got the hell out."

Over the next few months, Loomis and Thorne resigned from a succession of bank directorships and utility boards, including the Commonwealth & Southern Corp. and the Public Service Corporation of New Jersey. By Christmas of 1934, the only title Loomis retained was president of the recently formed American Superpower Corp. He and Thorne sold out virtually all their holdings with the exception of United Power, which was the one stock they would never part with. They sold all their shares in Bonbright and put the proceeds into a holding company that Loomis entrusted Thorne to manage as a personal investment company. Without so much as a backward look, Loomis quit Wall Street for good.

Loomis had discovered early in life where his talents lay and had twice set off in the direction of a more reliably profitable future, first in law and then in finance. His life had now come full cycle, with his enormous wealth enabling him to recapture his schoolboy's love of invention. "It was a counterattraction—he always liked the other girl better," observed William Golden, a wealthy investment banker who also turned to a career in science after World War II. "He was a very active, creative man. He had already been very successful in two different professions. He didn't just want to double his money and be the richest man in the graveyard. What would be the point in that?"

Loomis never once expressed any regret at leaving the business world. In some ways, he seemed to have no regard for money. "Having made it that way, and so quickly, it became impersonal with him," said Caryl Haskins. "He wasn't interested in it. When he needed it to buy things—new instrumentation, new technology—he bought them. That was all he really cared about. And of course, he gave a lot of it away. But he never talked about that with me. I don't think he thought his past accomplishments were worth mentioning."

Thorne continued to keep one foot in the financial world and ran the Thorne Loomis investment company, operating out of a large office in the law firm of White & Case at 14 Wall Street. He remained one of the largest shareholders in the First National and exercised considerable influence in banking affairs. He was already much wealthier than Loomis, and in the coming years his assets would far outstrip those of his brother-in-law. Thorne had always known that this day would come and had long strived to keep Loomis interested in their joint ventures, even if it had often meant shouldering more than his share of the work. He had felt fortunate to have someone of Loomis' brilliance as his business partner, if only on borrowed time. Although they went their separate ways, according to Henry Loomis, their partnership lasted to the end of their lives. "In a sense, they never really stopped working together. Alfred was always in the background. He just spent more time doing what he wanted to, which was physics."

There were some hurt feelings on Thorne's part that Loomis did not invite him to participate in his scientific work—"as if he were somehow not smart enough," said Evans. Loomis acted as though he were engaged in some exclusive activity at Tower House, and while this provoked some jealousy, nothing was ever said about it to her knowledge. Thorne was always very supportive of Loomis' scientific work and on more than one occasion contributed substantial sums to his research projects.

The changing climate on Wall Street undoubtedly hastened Loomis' departure. Fear and cynicism had turned the American public's mood ugly, and their favorite target was New York's banking elite. Loomis was shaken by the imputations of self-serving dishonesty. He had been among a small group of men with enormous private power, and now the public and Washington were holding them accountable and seeking to curb their influence. Like most of his fellow club members and board directors, he did not trust Roosevelt. He did not believe in initiatives like the Industrial Recovery Act, which put the state in partnership with big business, and he put his faith in private enterprise over politicians and their blundering bureaucracies any day of the week. He strongly believed the corporate form was by far the simplest and most efficient way to raise capital to build dams and power plants and modernize old industries. In his view, Washington could never have raised the money and planned the expansion of the public utility industry half as well as Bonbright and other investment companies had done in the twenties. But Loomis had no stomach for the endless infighting and no intention of spending his time colluding with Republican malcontents like Ogden Mills, "mounting assaults on the fortresses of the New Deal."

Loomis saw the Tennessee Valley Authority (TVA), established in May 1933, as a symbol of the worst of the New Dealers' tactics. "He thought it would destroy the business world," recalled Paulie Loomis. The TVA undertook a massive program of construction, engineering, and education to harness the Tennessee River, improve navigation, and supply low-cost electricity to the rural South. The utilities industry naturally resented this form of direct government competition and claimed that the TVA enjoyed an unfair advantage while they were being forced to reduce their rates. The TVA works program became such a source of bitter controversy that the law firm of Winthrop & Stimson was called on to file a nineteen-company suit on behalf of Commonwealth & Southern (the holding company whose board of directors had until only recently included Loomis and Thorne), which had interests in Tennessee and ten other states. The suit challenged the constitutionality of the TVA, which they argued was "the entering wedge" for the eventual "government ownership of all essential industries." The case was lost on the ground that the plaintiffs lacked the right to sue.

Loomis was appalled by all the litigation and wanted no part of the endless and complex regulatory proceedings. Stimson, who had returned to his law practice, became embroiled in related litigation and later was retained by Wendell Willkie, then president of Commonwealth & Southern, to negotiate the sale of one of its subsidiaries to the

TVA. For Loomis, though, the final affront came a few years later, when Congress held hearings on banking practices and Bonbright came under scrutiny for its huge profits. Though never accused of any wrongdoing, he adamantly refused to testify and defend himself against questions about the greedy design of big holding companies, which essentially allowed for an unlimited chain of acquisitions. The attack on his honor turned Loomis permanently away from politics and left him with the profound sense that it was a game whose rules he neither liked nor fully understood. He continued to support Stimson's political career loyally and hoped he would one day run for president. But more and more, Loomis felt a revulsion for public life and preferred the sanctuary of his laboratory. "He didn't want to have to fight the world," said Paulie Loomis. "He was happy to let Henry Stimson fight for him."

In the end, Thorne went to Washington alone, though he spoke for both of them when he preached strongly against too much government interference, warning, "Legislation which restricts the management of such companies unduly or limits their investments arbitrarily cannot fail to hurt the economic developments of the country."

"Alfred hated all that," said Ed Thorne. "He was very private, and he was not all that comfortable with controversy. He didn't want any public role. It was the only time I ever heard my father complain about him."

Loomis retreated to his scientific Valhalla in Tuxedo. His business associates were shocked by his self-chosen exile, which ended one of the bluest of blue-ribbon careers. From time to time, whenever a prestigious board seat came vacant, his name would be mentioned in connection with the plum job. When Loomis was honored by the American Association for the Advancement of Science in March 1933, a *New York Sunday* writer observed that his "growing fortune" had clearly become "secondary to his scientific preoccupations," and the transition from finance to physics, long in coming, was now complete: "He is chosen as one of 250 American scientists, listed as foremost in their different fields of research. He is a physicist."

As the *Journal* wistfully concluded a few years later in a story on new business leaders, there was little chance that Loomis and Thorne would ever be lured back to "active duty" on Wall Street: "Shortly after the depression started, they went into retirement with their fortunes. Thorne spending more time fishing and hunting and Loomis with scientific work at his place up the Hudson. Loomis is now one of the top ranking scientists of the country. . . ."

GIVEN their spectacular success, it was only natural that Loomis and Thorne extended their partnership to include many projects outside business, based no doubt on their compatibility, mutual confidence, and shared appetite for new challenges and adventures. Although the country was in a depression, they were both vastly enriched, emboldened, and in high spirits and began to indulge in many of the luxurious pastimes enjoyed by their fellow millionaires. "Landon was always the instigator in those years," said Paulie Loomis. "Landon could talk anybody into anything, and Alfred would always make it happen. They were made for each other, and they had a ball together."

While Loomis certainly enjoyed living well, and maintained two large mansions in Tuxedo staffed by a dozen or more servants, along with a twenty-room duplex penthouse at 21 East 79th Street off Fifth Avenue, his primary interest was comfort and convenience, not the pursuit of stylish social ostentation, which was always more in Thorne's line. Like many of New York's gilded aristocrats, Thorne liked everything on a grand scale. He went on gaudy buying sprees, collecting art, antiques, and horses, all of which came with fancy pedigrees and even fancier price tags. His magnificent 230-acre Long Island estate, which he named Thorneham, boasted its own indoor and outdoor swimming pools, indoor tennis court, athletic house, and stables. Thorne maintained several large trout ponds on the property, which he kept well stocked, so the fishing was always good. He commissioned the famous landscape architect Umberto Innocenti to design the estate's ornate gardens, which for many years were considered among the finest in the country.

"Landon was the showman," said Bart Loomis, Alfred's grandson, recalling the ostentatious $100,000 Tiffany sapphire necklace Thorne once presented to his wife, Julia. "He liked to own things—yachts, islands, railroads, you name it, he bought it. You want to talk about big money, Landon Ketchum Thorne was big money. He was the flashy one. That was not Alfred's way, but he went along with it."

Despite their contrasting personalities, Loomis and Thorne were as close as two men could be, professionally and personally, and Loomis enthusiastically joined in Thorne's extravagant hobbies and sporting pursuits. Loomis' three sons were close in age to Thorne's two boys, Landon junior and Edwin, and the two families became inseparable, visiting each other often and going off on extended holidays together. In large part this was due to Julia Thorne, who was as charming and charismatic as her husband, and Loomis doted on her. Compared to his sister, Loomis' wife, Ellen, seemed sweet but rather vague or, as one family member put it, "a bit on the fey side. At times, you wondered if she was really all there." "Julia was very overpowering and kind of spoiled," noted Paulie Loomis, "but both Alfred and Uncle Landon thought the world of her."

Julia was at the center of the two tribes, and she "loved to hold court." She threw splendid parties that were famous for the opulent gold dinner service adorning the table and the giant flagons of champagne that white-jacketed waiters tipped into glasses. Witty, intelligent, and unusually well read for a woman of her day, "she was one of the most informed women you could ever meet," said her daughter-in-law Mimi Thorne Gilpatric, who was married to Landon junior. "She could hold forth about anything that was going on, whether it was politics or literature, and she always had something interesting to say." She was a connoisseur of rare manuscripts and first editions, and assembled one of the foremost collections of William Blake, which she later gave to the Morgan Library. With age, she became very grand, and commissioned an elaborate genealogy of the Thorne family's blue-blooded ancestry, which she had bound in leather, with hand-engraved parchment pages and an elaborate gilded crest on the cover.

Julia was in every way "a glamorous figure," recalled Gilpatric. "She would be sitting there in one of her Mainbocher gowns, and all her jewelry, with five rings on each finger and gold bracelets from here to here"—pointing from wrist to elbow—"she looked like one of the royal family."

During the winter of 1930, Thorne had been regularly featured in the sporting columns as a member of a syndicate of yachting enthusiasts headed by Paul Hammond, a great sailor, who were building the *Whirlwind*, one of the four yachts that would compete for the honor of defending the America's Cup. The other three aspirants were the *Enterprise*, owned by Harold S. Vanderbilt; the *Weetamoe*, backed by a syndicate headed by Junius S. Morgan and George Nichols; and the *Yankee*, belonging to a Boston syndicate headed by John S. Lawrence and

Chandler Hovey, with Charles Francis Adams, secretary of the navy, sailing her. But as it turned out, Hammond quickly got in over his head—spending a bundle on a second large boat to house his crew of thirty—and halfway through construction informed Thorne the syndicate was busted. So Thorne, with his deep pockets, took control of the syndicate and, as he did in all things, brought in Loomis as his equal partner. The two already owned several boats together, and Loomis agreed to split the cost of completing the yacht.

The two partners were prepared to spend whatever it took to win the America's Cup. It was an unbelievably extravagant undertaking at the time, as J-class racing sloops were built exclusively by powerful syndicates of wealthy families such as the Astors and Vanderbilts and not by self-made men like Loomis and Thorne. While there is no record of how much money they lavished on the *Whirlwind*, a good measure lies in the ultimate fate of the J-boats themselves, which became too prohibitive even for the very rich, so that by the end of the 1940s, America's Cup races were relegated to smaller twelve-meter boats.

Of course, the *Whirlwind*, which took its name from a famous clipper ship owned by a Thorne ancestor, was destined to be no ordinary racing sloop. The boat featured a highly original, innovative—and untested—design, which Loomis and Thorne hoped would give them the edge in the competition and win them the right to race against the British, namely Sir Thomas Lipton, in one of his periodic attempts to capture the cup from the New York Yacht Club. As one critic marveled at the time, "No cup yacht or any other racing craft ever carried quite so many newfangled ideas, and anyone who has the opportunity should carefully examine this marvelous creation." The *Whirlwind*'s radical design was created by L. Francis Herreshoff, son of the famous Bristol naval architect Nat Herreshoff, who had turned out the previous five America's Cup defenders. Rumor had it that the father had lent a hand in the design, and the boat, which was being built at Lawley's shipyard in Boston, was the subject of many flattering notices and much fanfare, accompanied by intense speculation. "There has been much mystery about her," *The New York Times* reported in March 1930:

Until the Whirlwind is overboard and rigged it will be impossible to pass judgment on her beauty. She is the longest and largest of the four defense boats, stretching several feet more on deck than any of the others, and is 158 tons, thirteen more than Yankee, the second largest. . . . She is the only real cutter, depending entirely on her keel, whereas the others have centerboards to put them in the sloop class. Also she is the only one of the four to have wood planking. She is of composite construction, with mahogany over steel frames. . . . With her large hull the Whirlwind will not be able to spread as much sail as the other boats. If she did she would not keep within the measurement limitations. A larger, lighter hull with less canvas is the plan for her.

Although his sailing experience had been mostly in racing small boats, Thorne was determined to take the wheel, while Loomis, who was an expert navigator from his days on Long Island Sound, tackled that job with his usual acumen and unwavering self-confidence. Naturally, Loomis adopted a scientific approach and enlisted the help of MIT's Naval Architecture Department, and together they undertook a thorough study of hull shapes. During the test program, they also decided to replace the 158-foot mast with a metal one—at a cost of approximately $25,000—constructed of duralumin, the strongest and lightest aluminum alloy. In the preliminary races, the yacht still did not seem balanced, or "in her groove," and they thought if they stepped the mast twenty inches forward, it would improve her windward handling and make her faster. But it took two men tugging at the wheel to keep her sails full and drawing in a breeze, and one time a helmsman was thrown clear over the wheel. In an effort to correct the fault, they decided to move the mast up again, a total of five feet forward from its original position. The boat was hauled out of the water so many times that summer, and was so many weeks behind her three rivals in shaking out her sails, that the odds were heavily against her in the preliminary betting.

Loomis hoped to improve her chances by seeing to it that the *Whirlwind* was outfitted with any number of ingenious devices of his own design. At the time, people joked that you could always spot the Loomis-Thorne boat a mile off because of all the whirling gigs and wind trackers that covered the deck. "Of course, Uncle Alfred developed some navigating equipment that was kind of the last word, and later a lot of people adapted his ideas to other things," said Ed Thorne, who was sixteen at the time and split the job of cabin boy with his brother, taking turns reading the courses and speeds at ten-minute intervals for the navigator. During the trials, according to Ed Thorne, Loomis never once came above deck but remained below, surrounded by all his gadgets: "We had a couple of races on foggy days, those days when you couldn't even see more than a couple of hundred yards. And of course in those days, you didn't have radar. But with Uncle Alfred's calculations, we'd always find the mark, even when all the other people had trouble. The only race we ever won was because of his navigation skills. We cut the right buoy and everyone else cut the wrong one. It never even occurred to my father to come about that buoy, because Alfred said it was another fifteen seconds or thirty seconds, or whatever. They had complete trust in each other, they had that kind of partnership."

Ultimately, the *Whirlwind* proved to be something of a clinker and finished dead last in the final race off Newport, Rhode Island. The victor, Vanderbilt's *Enterprise*, later defeated Lipton's *Shamrock V*. The *Whirlwind* was the outstanding failure of the America's Cup that year: in twenty-two starts, she won only one race. The yacht, expected to be the one with the greatest power, never attained her potential and at times "looked as if she were trailing a sea anchor." Everyone agreed, however, that she had by far the most spacious and luxurious living quarters of any America's Cup defender. While some later carped that the yacht might have fared better if Thorne had allowed himself to be relieved at the wheel by a more experienced skipper, the larger problem—which would often prove to be the case with the younger Herreshoff's creations—was finding her proper trim and rig. But at the end of the season, when it came time to settle the expenses, which all hands knew had run nearly half a million dollars, Thorne forever endeared himself to his friends when he told them "to forget it," adding that he and Loomis "would look after the matter."

In 1931, finding Long Island too tame and increasingly suburban, Thorne saw an opportunity to buy Hilton Head Island, in South Carolina, which before the bridge was built was an isolated strip of beach reachable only by boat. The island paradise, once home to wealthy southern landowners and sprawling plantations, was occupied by Union forces during the Civil War, and on General William Sherman's orders, the confiscated land was sold to freed slaves for a dollar an acre. When the cotton crops failed, many abandoned Hilton Head, and the island was almost forgotten. In the 1890s, hunters began buying the property for recreational purposes, and a New Yorker named William P. Clyde eventually managed to acquire nine thousand acres, including the last of the antebellum houses, Honey Horn. When Thorne, whose ancestors had owned plantations in the Sea Islands for generations, heard that the current owner, a northern industrialist named Roy Rainey, had been ruined in the Crash, he proposed that he and Loomis buy the land and turn Hilton Head into a private hunting and fishing resort for family and friends. Loomis, who was by nature a loner, loved the idea of having his own island and was all for it.

"They paid about $120,000 in cash, because they had cash on the barrel head in those days," recalled Ed Thorne. "They bought up about twenty thousand acres, which was virtually the entire island. There was almost nothing on it, except what had been an old Confederate fort, and the Honey Horn plantation, which was a one-story house that dated back to the Civil War. They fixed that up and expanded on it, and they turned it into a marvelous sporting preserve. They built another boat together, the *Northern Light,* which they took there."

Even today, Betty Evans remembers being struck by how remote and fantastic the island was then: "There was no one there at all, only a few black families, and they spoke nothing but Gullah. It was the most beautiful place you've ever seen. Julia brought her horses there, and fixed up the house with the most elegant antiques and rugs. They threw huge house parties there, inviting all their family and friends. Early in the mornings, a boy would come and light a fire in the bedrooms while we were still asleep, and get the potbellied stove going in the bathrooms. Everybody would be up at the crack of dawn to go hunting. And you never saw such hunting. It had every animal known to man."

Loomis and Thorne hired a local islander, Mose Hudson, to patrol their property on horseback and keep trespassing hunters away, and to make sure the reserve was always well stocked with mink and other game. "This was hunting in a grand style, with horses and dogs, and elegant picnics," said Frederick Hack Jr., who grew up on the island and whose family later bought Honey Horn in 1950. "The whole plantation more or less operated on that level. It was quite something." After a typical morning shoot, the party would find a shady spot and set up camp. "It was just like on safari," recalled Paulie Loomis. "A black man would come along, all trimmed out in a waiter's uniform, and bring tables, and tablecloths, and chairs, and china. You couldn't believe your eyes. We'd have this fancy lunch outside, and they would grill quail, and great big oysters in the shell and melt butter. It was a feast."

In the evenings, everyone would dress for dinner, and Julia would arrange for a wonderful feast in the big dining room. There would be a roaring fire afterward, and Loomis would keep the boys busy by giving them mathematical problems to work out or challenging them to a game of chess. Loomis, of course, always played with his back to the board and often played several matches at once, all the while keeping up a lively conversation with his other guests. Sometimes he could be persuaded to put on a magic show, and he would mesmerize them for hours doing tricks with a quarter and a few matchsticks. At the end of the evening, they always brought out a bottle of what Thorne called "sippin' whiskey." The island was littered with illegal stills, and he and Loomis took great pleasure in locating them all and helping themselves. Added Evans: "It was 120-proof bourbon—it only took one sip."

At Honey Horn, Loomis enjoyed playing the country squire. He re-created Highhold, Stimson's gentleman-farmer's estate on Long Island where he had spent his happiest times as a boy. Over the years, with the help of a devoted manager named Ted Armstrong, he and Thorne transformed Honey Horn into a working farm, adding stables, servants quarters, a guesthouse, and laundry room. It was a totally self-sufficient compound, with its own milk cows, chickens, and a large vegetable garden. The generous offerings were described by a visitor:

> The big kitchen at Honey Horn sent out a tempting fragrance of roast turkey and venison, of duck with orange sauce made from bittersweet island oranges, Carolina shrimp pie, oyster stewed with crisp bacon and onions and served with fluffy rice. There was crunchy benne seed candy in the crystal dish, or perhaps a plate of pecan pralines, with the nuts fresh and crisp from island trees.

Having made their fortune in rural electrification, Loomis and Throne fittingly imported an electric generator, bringing the first power to the island. Compared to the simple farms kept by the neighboring black families, Honey Horn was a mecca of modern engineering: the mere fact that they had the only tractor had a huge impact on the local economy. Over the years, Loomis and Thorne bought land from any families willing to sell, and by 1936 the black population on the island dropped to only three hundred as compared with three thousand forty years earlier. The northern conquerors even managed to acquire the last large lot, the 803-acre old Confederate Fort Walker site, from the federal government, for an additional $12,600. They had the island virtually to themselves until World War II, when the turn-of-the-century lighthouse in Palmetto Dunes became the site of a marine encampment. Gun placements, for target practice over the Atlantic Ocean, were set up on the beach, and Loomis' sons remember collecting the shell casings that washed up on shore.

Over every winter for almost twenty years, Loomis and Thorne, along with their five sons, hunted to the hounds, shot skeet, fished, sailed, and entertained friends and important business executives, politicians, and a growing number of Loomis' scientific colleagues. Stimson came to stay, as did Ernest Lawrence. On one occasion, they even played host to the king of Sweden. Guests would take the train from Penn Station to Hardeeville, a small station on the Atlantic coast line about twenty miles north of Savannah, where they would be met at the station and brought over to the island on a small boat. One frequent visitor was Karl Compton, who together with Loomis spent countless hours digging up the beaches in search of old Indian burial grounds, arrowheads, and other artifacts, all to no avail. In the process, they discovered that the place was teeming with poisonous snakes—huge rattlesnakes, water moccasins, and copperheads. Loomis had Compton call on his experts from MIT. "They would come over and trap them, so they could extract the venom to make serum," said Evans. "I think Alfred also saw to it that some were sent to zoos."

Loomis and Thorne loved the island and collected all the information they could on its early history and the wealthy men who had come before them and built the "Big Houses," whose haunted ruins lay covered by weeds. There were also practical business reasons for becoming steeped in Hilton Head's past. Eventually the land would be divided, sold, and developed, and prospective buyers would want a clear title and the names of the previous owners and dates of sale. They faced only one problem: Beaufort County's land records had twice been destroyed, and no complete official history existed. Loomis and Thorne decided to fill in the gaps themselves and began researching the previous residents, their lots, titles, and sales, compiling detailed records of their own. An artist was commissioned to create a map showing the antebellum plantations, and the illustrated drawing was framed and hung in the hall at Honey Horn.

Loomis dedicated himself to making a scientific survey of the island, enlisting Compton's help on more than one occasion, and over the years he had a number of elaborate maps drawn up. He also attempted to give a full account of the wildlife on their island paradise, noting for the record that "as many as 4,000 Widgeon have fed in the duck ponds during a winter":

The island is used entirely for sporting purposes, ie:
Quail
Snipe
Dove
Duck
Deer
Wild Hog
Wild Turkey
Coon and Possum
   Fishing (Sea-trout, Bass, etc. in the rivers and creeks;) (Blue fish, Red Snappers, etc., in the ocean.) (The oysters are plentiful and unusually good—also crab and shrimp.)
   The covies of quail found in 1936 were 293—about 1,000 quail are killed during a season, the limit per gun a day being 10. The duck shooting is extraordinarily fine, and no less than 19 varieties have been seen on the Island, ie:

Mallard	Redhead
Black Duck	Greater Scaup
Pintail	Lesser Scaup
Widgeon	Ring Neck
Gadwall	Ruddy Duck
Blue Wing Teal	Bufflehead
Green Wing Teal	Whistler
Wood Duck	Hooded Werganser
Shoveller	Red-Breasted Werganser
Canvasback	Canada Goose

Much as he enjoyed Honey Horn, Loomis was probably happiest at Tower House, tinkering with the latest invention in his laboratory, listening to updates of the various experiments under way, and exchanging news and ideas with other scientists. But over the next few years, the island provided a welcome respite from his increasingly strained home life in Tuxedo. While they were the last family in the world to discuss

such matters openly, it had become painfully obvious that Loomis regarded his wife with something more akin to forbearance than affection. "She was a very dependent person, on her parents, on Alfred, and
on all the help," said Paulie Loomis. "She used to have people waiting
on her hand and foot, which Alfred just hated. He used to complain
that he couldn't walk into a room without bumping into one of her
'Irish biddies.' The fact is, I don't think they ever really got along."

Ed Thorne was not surprised that they drifted apart. "She was a
lovely, sweet person, but very, very Boston. She was not a gregarious
person at all, and sort of buried herself in her books and nineteenth-
century novels. It almost seemed as if she were kind of out of this world."
Over the next few years, Ellen, who was always becomingly "delicate" in
the Victorian sense, would suffer from recurring attacks of migraine
headaches and mysterious ailments and took to her bed for weeks at a
time. Loomis, following a practice that was then common to the point
of being fashionable in the upper and middle classes, packed her off to
various hospices and clinics for extended stays to be treated for nervous
exhaustion. In February 1932, during one of these confinements, Stimson wrote to Ellen in the hospital, and in typical WASP fashion advised
her to keep a stiff upper lip until she could regain her strength:

My dear little Ellen,
    When I remember how faithfully you used to write letters to me
in periods of stress and trouble, it makes me feel remorseful that I
have not written to you before. I did not like to write a dictated
letter, but I have been unable to get the time to do anything else. So
finally I have curbed my pride, and I am sending you this note of
affectionate encouragement.
    I am dreadfully sorry that you are having such a long tie-up, but I
am very glad that they are at least making you rest and avoid a little
of the terrific strain you have been under. I know how your heart
must be torn at being shut up and away from Dedham [her parents'
home in Massachusetts, where she often stayed when "undone" by
life and illness], but I also know that you will reconcile yourself
courageously to this as the best way to solve the problem and get fit
again. Your life is so invaluable to so many people that you must take
every step possible to save your health. If you can, send me a message
through Alfred sometime just to let me know how you are feeling
and getting on.

The exact nature of Ellen's complaints, and whether or not many of her symptoms might fall under the heading of "neurasthenia"—the term then given to problems that occupied the gray area between medical and mental illness—remains a matter of some debate within the family. "Oh, she was a terrible hypochondriac," said Evans. "The doctors prescribed all kinds of things, and she'd take some draft of medicine and go to bed for the day. If you went in to see her in the afternoons, she was always lying on the chaise in her room with the lights dimmed, the folds of her fancy lace gown perfectly arranged all around her."

But to many in the family, Ellen's illness seemed to serve as a convenient excuse for Loomis to pack her off, leaving him free to pursue his interests without distraction. "He kept trying to get rid of her," said Paulie Loomis. "She wasn't sickly, she was sick. But he wasn't very sympathetic. Alfred wasn't the kind of person who tried to understand people. He pushed a lot on her. And he changed. I think he changed into who he wanted to be and didn't have a lot of time for her anymore."

Loomis was not any easier a father than he was a husband. He set the bar very high when it came to his three sons. After he cashed out of Bonbright, he awarded each of the three boys a substantial share of their inheritance—roughly $1 million—on the theory that it was never too early to begin charting one's own course. The youngest, Henry, was only fourteen when he was given complete financial independence. "Father decided to divide the estate because he didn't want his money to become a tool of discipline with us boys," he once recalled. "I remember that his cronies were aghast when he gave us the money and warned that we would make all kinds of mistakes with it. Father said we would make mistakes inevitably, but he'd rather have us make mistakes with the sums we'd be playing around with as boys than the ones we might make if we had to wait until his death to get the money, when we were thirty or forty years old."

While Loomis routinely drafted his sons as "guinea pigs" in his research, he remained, for the most part, a distant and detached figure. They loved working with him in his laboratory and were fascinated by his projects and famous guests, but there was never any question that he was far more interested and engaged in his world than in theirs. As one of the many small rebellions that teenagers specialize in, Loomis' sons regularly played a rather hazardous game of chicken on the railroad tracks that ran through Tuxedo, competing to see which of them dared stand the longest in the path of the approaching express trains. Not even this deadly sport succeeded in attracting his attention, and according to Henry, Loomis never bothered to reprimand them: "We were never consciously influenced by Father in anything—about getting cars, going out with girls, traveling or staying home, where to go to college, or anything else."

Loomis was nothing if not difficult to impress, and his sons devoted much of their lives to trying to win his approval. They were fiercely competitive and went to great lengths to exceed their father's expectations—engaging in daredevil sports, immediately volunteering for duty in World War II, and later endeavoring to achieve a measure of his success in either science, business, or politics. They would spend months mastering complicated chess strategies at boarding school, only to come home to be cut down by Loomis in a swift series of moves. There was a merciless quality to his reason that always seemed to defeat them. Henry never forgot a remark his father made over Christmas in 1935, when he and his two older brothers were making plans for their summer vacation. "Lee and I were talking about crossing the Atlantic in a thirty-five-foot boat, and Farney was planning to climb mountains in India," he recalled. "Mother asked Father: 'Will you still believe in your theories about children if all three of them get killed this summer?' Father replied: 'Three is not a sufficient number to prove any scientific theory.' "

"Alfred was a wonderful man in his own way, but he wasn't really very much of a human being," observed Paulie, who felt her former husband, Henry, and his brothers had been treated to a "funny, hard sort of childhood. . . . Alfred was selfish to the point where he never really gave very much unless he enjoyed what he was doing. Everything was always calculated—what could be gained, what could be lost. As long as he was happy, that was all that mattered."

LOOMIS used his new freedom and financial resources to extend his largesse beyond his own laboratory to a great variety of philanthropic endeavors. Since the turn of the century, Andrew Carnegie had made funding schools, laboratories, and other institutions for the advancement of knowledge almost an obligation for newly rich American tycoons, particularly for those eager to put to rest any lingering doubts about how their money was made. As the billionaire who spent a lifetime building a steel empire had once stipulated, "Amassing of wealth is one of the worst species of idolatry." It was a Gilded Age sense of social responsibility Loomis' own grandfather had deeply felt, and he had dedicated his latter years to building tuberculosis sanitariums and medical schools. Now, at this low ebb in the country's fortunes, having accumulated an embarrassment of riches, Loomis found new purpose in giving away his money. He would use his wealth to encourage scientific investigation and provide opportunities for excellent men, while indulging in his favorite hobby at the same time.

He gave money to any project—no matter how large or small or whimsical—that appealed to his Baconian belief that scientific discovery could be claimed by anyone "keeping the eye steadily fixed upon the facts of nature." He himself was living proof of this principal, as he regularly became expert in a new field in only a few months, each time making important observations in the frontiers of astronomy, biology, and chemistry before moving on to the next of nature's challenges. Loomis possessed a fundamental American bias in favor of experimental rather than theoretical science, which was repeatedly rewarded by the pioneering research carried out by the succession of brilliant young researchers he attracted to Tower House. His modus operandi was simple: he furnished the men and the money necessary to attack a problem, and once they had made a contribution, he let the momentum carry the others forward while he hunted for new, more fertile fields of inquiry. As his son Henry once observed, "As soon as he starts working on the next decimal place in the final answer, he shoves off."

As a consequence of his insatiable curiosity, and impatience to make his reputation as quickly as possible, Loomis pursued a peculiarly eclectic strategy, attempting to advance and promote research in a wide range of disciplines, both through his own efforts and by funding the work of others. A great believer in self-education, Loomis, with Thorne's help, established a foundation to foster America's budding engineers, offering stipends so students could go on six-week tours of the leading industrial plants of the East and Middle West. He wanted to give young men who had not had his advantages of higher education an opportunity to travel and to see firsthand what was happening in modern manufacturing enterprises. The all-expense-paid camping trips sent ten boys at a time across the country in special Loomis-designed, gadget-laden Ford motor trucks that transformed into large tents, complete with sleeping quarters, kitchen, and "parlor."

The only requirement was that participants had enough money to pay for their own boots—a paper package containing a khaki uniform was provided free of charge—and agreed to keep a diary of everything they did and saw during their travels. Loomis contacted Compton at MIT, who agreed that it seemed "an interesting venture" and promised to send over enough "technology men" to fill one of the trucks. In all, two thousand young "industrialists" made the trip from Tuxedo to Kansas and back, and Loomis paid to have selections from their diaries printed privately in booklet form.

"It was a very extravagant trip," remembered John Modder, who like many local boys at the time considered signing up. "The average worker was making about a dollar a day, maybe a dollar and a half. Loomis was giving people sixty dollars to go on the trip, which would be like a thousand now. It was mostly college students, but anybody could go, and quite a few from Tuxedo did. And if you had some sort of idea you wanted to work on, you could go up there and tell Mr. Loomis, and he'd feed you and house you and everything else while you were working."

One newspaper columnist, noting Loomis' propensity to underwrite all manner of projects, predicted that if he kept up this hobby, his name would soon be "not so well known in Wall Street as he is in the field of science":

> It is in the field of physics that this studious financier has his primary interest. In Tuxedo Park, his home, he has built the Loomis Laboratories and any scientist who wishes to devote a week to a year of study to pure science (commercialism is as far from the minds of the true scientist as it is from the heart of a Rembrandt) is welcome to remain there as Mr. Loomis' guest. Mr. Loomis likewise is conducting his own studies, and he has surrounded himself with interesting scientific students, largely drawn from the faculties of leading American universities. He would rather sit in on an argument over the Einstein theory than tune in on an Aldrich versus Eccles debate [Mariner Eccles, head of the Federal Reserve Board] on excess reserves. . . .

Loomis was not interested in public acclaim so much as a certain kind of respectability, and many of his large charitable donations and anonymous gifts were clearly designed to court favor with the scientific establishment, which might otherwise have been inclined to dismiss him as a dilettante. There had long been a prejudice against "amateurs" in American science—the scores of laymen, collectors, and quacks who supported the many popular journals and publications but otherwise gave the profession a bad name. As far back as the 1840s, the physicist Joseph Henry had complained, "Every man who can . . . exhibit a few experiments to a class of young ladies is called a scientist." Since that time, Henry and his allies at Harvard and other institutions of higher learning had organized and closed rank to keep amateurs from playing a prominent role in research, and to that end, they created a national professional organization, the American Association for the Advancement of Science, a fraternity over whose membership they could guarantee control. Even more exclusive was the National Academy of Sciences, a private organization created during the Civil War to provide advice to the government and one that Loomis longed to join.

Aware of their deep-seated prejudice against "gentleman scientists," Loomis launched a calculated campaign to win over influential members of the profession. He was willing to pay his dues and to earn membership in their club. His famous laboratory gave him a strategic advantage, and he sponsored dozens of scientific conferences, meetings, lectures, and dinners, increasingly ingratiating himself with cash-poor universities and scientific agencies. Loomis, along with Wood, sat on the planning committee to select science exhibits for the 1933 Chicago World's Fair, and several of the meetings were held at Tuxedo Park. The fair's triumphant theme that year was "the Century of Progress," highlighting the benefits of science and technology. Unfortunately, it was the midst of the Depression, a time when many politicians and leading public figures were questioning what the industrial advances had wrought on society, so Loomis reached into his own deep pockets to fund many of the large-scale exhibits—doing pro bono public relations work for his chosen field.

With his dedication to pure research, willingness to play favorites, and unlimited funds, Loomis turned himself into a powerful behind-the-scenes force. Along the way, he made some important friends, chief among them the popular Compton. The two men were a natural fit: Compton with his genial manner and political skill, Loomis with his wide-ranging intellectual curiosity and talent for fostering other scientists' work. Thanks in part to Loomis' maneuvering, Compton was offered the job as MIT's new president in 1933. As a testimony to his loyalty, Compton in turn stated as one of his conditions to taking the job that his influential patron be named a trustee. According to Loomis, it was not an empty bluff: "As a new president, he wanted to know if he asked for things, he could get them. He probably used this as a test to see how serious they were," he said, adding, "They wanted young blood. He was young and I was young." He would go on to help establish MIT's graduate school, providing substantial financial aid and raising funds from other business leaders.

As adept as he was at playing politics, Loomis never underwrote any project that did not appeal to his own specific studies and interests. During the height of the Depression, when most academic journals teetered on the edge of bankruptcy, he offered to foot the bill for the fees assessed by the *Physical Review,* so that new and interesting work could be published without regard to cost. For years afterward, anyone who submitted an article received a bill along with a form letter from the *Physical Review* that stated: "In the event that the author or the institution is unable to pay the page charges, these will be paid by an anonymous friend. . . ." While his identity as the *Review*'s "angel" was not made public until after his death, it was common knowledge among the presidents of leading institutions, who in turn began appointing Loomis to their prestigious boards and committees.

By June 1933, when Yale University conferred the honorary degree of master of science on its alumnus, Loomis was lauded as a man who defied traditional categories and whose experimental approach made him a bold example for the times. The citation listed his several identities— "lawyer, businessman, physicist, inventor, philanthropist"—and compared him to the prototypical American physicist: "In his varied interests, his powers of invention, and his services to his fellow-man, Mr. Loomis is the twentieth century Benjamin Franklin."

All his efforts made certain that while he had retired to Tuxedo Park, it would not be for long. He was only forty-five years old, and his finest work was still ahead of him. His reputation—carefully crafted by himself and the wily Robert Wood—had spread to Europe, and Tower House had gained international fame as one of the best-equipped and most interesting private American laboratories. Whereas today physicists look forward to spending August studying at a school on Lake Como, in the 1930s the lucky few received an invitation to spend the summer at Tuxedo Park, working with the famous scientists who gathered there, including Fermi, Bohr, Heisenberg, and Einstein.

Alvarez, who was working in the Berkeley Physics Department, first heard of the legendary Alfred Loomis from a Berkeley colleague named Francis Jenkins, who had spent a summer at Tower House as Loomis' guest. "He had told me in wide-eyed amazement about the fantastic laboratory in Tuxedo Park, and about the mysterious millionaire physicist who owned it," recalled Alvarez. Jenkins told him in confidence that he was sure Loomis was the *Physical Review*'s anonymous benefactor. He thought Loomis a "wonderful person" but made it clear he did not care for the other residents of the park. "He thought they were too 'snooty,' and looked down on the scientists as barbarians who 'didn't even dress for dinner.' "

Chapter 6

RESTLESS ENERGY

He was interrupted by the entrance of a tall and well-rounded blonde of highly unscientific appearance.

—WR, from *Brain Waves and Death*

ANY visitor to Tower House in the mid-1930s eventually had to submit to the ritual known as "putting on the electrodes," which was part of the preparation for the measurement of brain waves, the research that would consume Loomis for the next few years. Guests who thought they were being taken on a guided tour of the laboratory would suddenly find themselves being eased into a chair as Loomis cheerfully talked them into undergoing a few harmless tests. Eminent biologists, chemists, physicists, psychiatrists, and neurologists, along with their wives and any other houseguests who happened to be stopping overnight in Tuxedo, were all recruited as volunteers for his experiments.

The preparation itself was not so much painful as disconcerting. It consisted of snipping a few hairs at various places on the scalp, rubbing the exposed skin with ointment to facilitate electrical contact, and attaching small silver disks called electrodes. One electrode was placed high on the forehead, one on the crown, and the third on the occiput, or back of the skull, so the action of the front part of the brain could be registered separately from that of the rear. (After many complaints about the lab's amateur barbers, Loomis and his colleagues found they could "obtain satisfactory electrical and mechanical contact with the scalp without even cutting a hair.") Either way, the result looked quite frightful, with the flexible wires previously soldered to the electrodes hanging Medusa-like from the subject's head so they could later be plugged into connections to the measuring apparatus in the control room.

Loomis would then briskly march his victim to one of the downstairs laboratories, which had been converted into a "sleeping room," where they would be asked to take a supervised nap. Richards, who collaborated with Loomis on this project as on others, wrote about the experiments in great detail in his novel, *Brain Waves and Death*. In an author's note in the front of the book, he explained that electroencephalography was established as a science in 1933 and asserted that the science part of his fiction was wholly accurate. "Far from having no resemblance to anything whatsoever, the facts here given about it represent our present knowledge of the subject, and are false in so far as that knowledge is incomplete." What follows is his description of the "sleeping room" at Tower House, which corresponds to the more clinical account featured in Loomis' paper "Electrical Potentials of the Human Brain," published in the *Journal of Experimental Psychology*:

It was a cheerful, undistinguished little place, much like hundreds of others in country houses up and down the East Coast. Care had been taken to conceal as much as possible that it was part of a scientific laboratory, for nervous people do not like to be surrounded by electrical instruments. It contained, besides the couch on which the body rested, a comfortable chair, a built-in wardrobe cabinet of modern design, and a large bed with small tables at each side of its head. On the wall were a couple of sporting prints and the curtains framing the shuttered window were gay. Only a sort of gigantic wall-plug above the head of the bed, and three search lights set in the ceiling, were at all unusual. . . .

The searchlights were a source of infrared light. A camera was discreetly mounted in the wardrobe to provide a photographic record and document changes of position. As the rooms were darkened while the subject was sleeping, Loomis used cameras with the fastest lenses then available and a new type of infrared film that the Eastman-Kodak Company made up for him specially and reportedly had flown to Tuxedo Park from Rochester on dry ice when he was doing experiments. The laboratory was equipped with two professional-quality darkrooms where the film was developed.

The wires from the subject's head were plugged into the wall socket above the bed. The sleeping room was in the far corner of the laboratory, secluded from everything, and soundproofed with copper netting to eliminate any interfering electrical noise. A microphone near the bed was connected to the second so-called amplifier room, which contained the high-fidelity amplifiers Loomis used to magnify the minute electric impulses given off by the brain, making them much stronger, in the same way radio signals are amplified in a receiver. According to Richards, it was packed with black metal filing cabinets filled with radio tubes and meters and "resembled an overgrown radio transmitting station." The amplifiers were also used in recording heart rate, bed movement, respiration, and sound communication between the sleeping room and the third, or control, room sixty-six feet away.

In the control room, a loudspeaker that was connected to the microphone in the sleeping room picked up even the slightest noise or sigh the subject might make. When they turned it up, the rustle of a sheet sounded "like a forest fire." Loomis used three amplifiers, so that signals from each part of the head could be recorded simultaneously. Each amplifier was attached to an electronic pen of silver tubing, which recorded the impulses or brain waves on a continuous ribbon of paper on a huge revolving drum. One pen wrote in red and the other two in blue ink. The forty-five-inch round drum, eight feet long, spun around once every minute, each second of "brain current" leaving a wave three-fourths of an inch long. Seven hours' sleep stretched out to a wave line 1,575 feet long. There was also an oscillator for sending tones of various frequencies to stimulate the subject in the sleeping room. Garret Hobart, who was an amateur radio enthusiast, manned the control room and kept Loomis' complicated custom-rigged apparatus running smoothly. As all three rooms were soundproofed, making communication between them difficult, Loomis installed an elaborate intercom system like that used in many law offices at the time. Thus his various itinerant experimenters, who generally included Richards, Hobart, and Loomis' longtime collaborator E. Newton Harvey and his wife, Edith, could monitor the experiment no matter where they happened to be. For the most part, however, they just shouted back and forth.

Loomis and his co-workers found that when the subject was awake, the pens would draw little trains of perfectly symmetrical waves, but as he fell asleep, there would be bursts of faster waves interspersed with little activity, when the pens would move sluggishly, dragging jagged spikes. As the subject fell into a deep sleep, the waves became larger and appeared more frequently. The ink tracing made on the paper was known informally as an "afternoon sleep record," because the experiments were routinely performed during an after-lunch siesta (though as their studies progressed, they began monitoring their subjects during a whole night's sleep). Loomis' earliest subjects included many family members. The Thornes, both their boys, and all three of Loomis' sons heroically endured being wired to the complicated apparatus for hours at a time so that precise electrical measurements of their brain waves could be made while they were sleeping. Afterward they were rewarded with much laughter and teasing about the gargantuan loudness of their snoring. "Aunt Julia said Alfred was always sticking those things to her scalp, and apparently it itched terribly," recalled Paulie Loomis. "She loved her brother and would do anything to make him happy. She would sit there for hours while he did experiments on her."

Loomis had first become interested in the new discovery of brain waves in the early 1930s after reading about the work of Hans Berger, a German psychiatrist. Berger had published his observations of the rhythmic electrical output of the human brain in 1929, but American physiologists had been unable to duplicate his results, and many were inclined to doubt the existence of the low-voltage signals he described. Loomis was intrigued and wanted to see if he could replicate Berger's observations using an electronic apparatus of his own design. In June 1935, in the first in a series of pioneering studies of brain waves, Loomis excitedly reported his findings in the weekly journal *Science:*

Recent interest in brain potentials has induced us to put on record the results of experiments carried out in the Loomis Laboratory, Tuxedo Park, in which the new phenomenon in this fascinating field has appeared most clearly—namely, the very definite occurrence of trains of rhythmic potential changes as a result of sounds heard by a human subject during sleep. . . .

Loomis, working with Hobart and Harvey, continued to investigate different aspects of these "trains" of waves, also known as Berger rhythms, which appeared in most adults and were regular rhythms with definite frequencies, usually between nine and eleven cycles per second. A few months later, Loomis reported the preliminary results of experiments performed on eleven human subjects ranging in age from five to forty-eight years old. They had identified four clearly defined states of sleep, or "levels of consciousness," occurring in the human brain, which produced different wave patterns, that they named according to their appearance, "spindles, trains, saw toothed, and random." (He later amended this finding to include five "levels of consciousness.") The brain wave patterns during sleep were so characteristic, Loomis reported, they could be used as a criterion to determine the subject's state or depth of sleep. The waves changed shape when a drowsing subject was roused, spoken to, or became restless. "In animals (dogs, rats) types of waves appeared quite different from any observed in human subjects." Deep anesthesia stopped the waves in a rat completely, and they did not start up again until the anesthesia began to wear off.

Loomis proceeded to investigate the pattern of brain waves of hypnotized subjects. He had been fascinated by the art and science of hypnotism since he was a boy and had first seen magicians achieve the strange transformation with the swinging action of a pocket watch. He had kept up a lively interest in the subject, and this gave him a chance to gain firsthand experience. For this series of experiments, Loomis recruited David Slight, a doctor from McGill University in Montreal who specialized in hypnosis. Slight brought with him as his "subject" a ship's carpenter, who was forty-four years old and had been hypnotized many times. He was first tested awake and then asleep and showed characteristically normal trains of waves. Loomis described the surprising results in *Science:*

> After Dr. Slight had induced the hypnotic state, a sustained condition of cataleptic rigidity ensued. Nevertheless, the trains characteristic of a person awake remained at all times during hypnosis and no spindles or random waves (characteristic of normal sleep) appeared during any of the tests.

Although he was thoroughly hypnotized, his brain waves were just like those of a normal waking person. No wave of the kind associated with sleep was elicited. Loomis concluded, "It would seem that the term hypnotic 'sleep' is not a correct one for the hypnotic state, at least not measured by this criterion."

Loomis wanted to see if the hypnotized person was subject to suggestion. He designed an experiment that would prove if it was possible to produce and record temporary blindness induced by hypnotic suggestion. The subject's eyelids were fastened open with adhesive tape and he was hypnotized, lying rigidly on the bed and staring fixedly. After giving the subject careful instructions, Dr. Slight suggested alternately, "You can see," and fifteen seconds later, "You cannot see—you are blind." This procedure was repeated "a large number of times," according to Loomis, and "in every case trains appeared when the suggestion was made that he was *blind,* and in every case they ceased when the suggestion was made that he could *see.* This was true both when there was a light in the room and when the room was in total darkness." This enabled them to establish that temporary blindness could be produced by hypnotic suggestion.

In a subsequent experiment, Loomis tested for the effects of emotional disturbance, such as embarrassment or apprehension. His youngest son, Henry, who was sixteen years old at the time, proved to be a particularly good subject for hypnotic suggestion. He recalled that for one experiment, he went to the sleeping chamber and immediately dozed off. His father then whispered in his ear, through a tube, that the *Lands' End,* the forty-foot sailboat he and his brother Lee had built, "was on fire." Henry remembered that he "jumped up" and then "started to climb the companionway ladder—but, of course, there was no ladder." He was still sound asleep, but the very suggestion that his boat was in danger stopped the normal brain rhythms. The experiment showed that while purely mental or intellectual activity had no effect on brain rhythms, emotional disturbance profoundly altered the wave trains. Loomis was so pleased with the result, he used it as an example in his paper on the subject, though Henry is identified in the paper only as the teenage subject H.L.

To relieve the tedium associated with the afternoon and nighttime sleep records, which required hours of monitoring the electric pens as they drew waves on the continuous paper, the experimenters resorted to all kinds of practical jokes and schoolboy pranks. One of the principal mischief makers was Richards, who, according to Hobart, had "a great sense of humor and was always pulling some sort of gag on someone."

Richards used to complain that Loomis' EEG apparatus was alarmingly similar to the "technique of electrocution" employed at Sing Sing and always claimed he had to have a drink or two to alleviate both the anxiety and the boredom the contraption inspired. The latter was evidenced by at least one of Loomis' experimental results on the brain waves on an unidentified subject during "an alcoholic stupor." The subject had "consumed 500cc. of gin (212 cc. pure alcohol) within 30 minutes," Loomis wrote, and "showed a marked large alpha rhythm with secondary potentials on the regular rhythm quite different from the rhythm which had appeared before the alcohol was taken. At this stage the frequency of the alpha rhythm was very definitely slower . . . the marked alpha rhythm only began to disappear when the subject had been deeply asleep for one half hour." Loomis concluded, "Alcoholic stupor like hypnosis does not exhibit the characteristic of sleep."

The local papers gleefully lampooned what they regarded as Loomis' strangest obsession yet. "Now picture the latest experiments at the laboratory," reported the *New York Evening Journal* on March 6, 1936. "Dr. Slight does a Svengali. His subject loses sensation and will power. His muscles become rigid. He's hypnotized. But his 'brain electricity,' tapped again, fails to register the rhythm of sleep. . . ." Thanks to Loomis, it had been proven once and for all that "the hypnotist's 'hocus pocus abracadabra' doesn't really put a person to sleep."

While Loomis' brain wave experiments were made light of in the popular press, his research was being taken very seriously in academic journals. His electroencephalograph was of a novel and highly efficient design, and his results were having a profound influence on the field, which was still in its infancy. Addressing a packed audience at the annual meeting of the National Academy of Sciences in the spring of 1937, Loomis revealed that his most recent experiments indicated that the human brain came equipped with an automatic "electric clock." Loomis described the "time clock" as a "subconscious cyclic process of some sort going on in the brain, which is no doubt the basis of our perception of time intervals." It was this subconscious rhythm, according to Loomis, that offered the first scientific explanation for an ancient puzzle: the ability of some individuals to wake from a sound sleep at a given hour, which they had fixed in their mind before retiring.

Loomis, along with Harvey and Hobart, made the discovery while studying the electrical brain waves "of a very sleepy yet conscientious person trying to obey instructions." The subject was told to open his eyes when a tone was sounded in his ear and a light signal was flashed, then to close them again. The signal lasted five seconds and was repeated automatically every thirty seconds. Eventually the subject grew accustomed to the noise and fell asleep. But a flurry of anticipatory waves shot through his brain two and a half seconds before each sounding of the tone. Although asleep, the subject knew the tone was coming, which according to the experimenters was evidence that the human brain came equipped with its own automatic "electric clock."

At the same time Loomis was doing his pioneering research in Tuxedo Park, Hallowell Davis, a professor at Harvard Medical School, was also conducting experiments with the electroencephalograph. In 1934, Davis had failed to detect the rhythmic electrical output of the brain and had initially discounted Berger's observation as an artifact. But when two of his students were testing his equipment on him, they observed Davis' own unmistakable alpha rhythm. Davis was the first to replicate Berger's findings west of the Atlantic. Only a few months later, Loomis successfully tapped the electrical output of the brain in a second set of experiments at his laboratory. Davis had immediately recognized the clinical potential of the electroencephalograph (EEG) and in collaboration with other researchers soon identified the characteristic wave pattern of petit mal epilepsy. After hearing about their research, Loomis invited Davis and his wife, Pauline, to Tuxedo Park and offered to fund further studies investigating the clinical applications of the EEG.

Between 1937 and 1939, Loomis, working together with the Davises, made major advances in relating EEG "disturbance patterns" to various "levels of consciousness"—from emotional tension and mental activity to relaxation and different stages of sleep. During these experiments, Pauline Davis was the first to observe cortical electrical responses to auditory, visual, and somatosensory stimuli in waking subjects. As Loomis was now devoting himself to scientific work full-time, his name appears as active collaborator in nearly a dozen papers on electroencephalography published by the laboratory in those years, and several of them were of great importance. In the end, Loomis' experiments succeeded in validating Berger's discovery, which became the basis of a new branch of study. He played a major role in the development of the electroencephalograph, which went on to become an extremely valuable diagnostic tool and is used routinely in hospitals to detect epilepsy as well as many other diseases.

At the time, it was believed that the electroencephalograph might be applied to psychological analysis, and Loomis invited an array of distinguished psychiatrists and physicians to Tower House to observe their work. His experiments had shown that hypnotism was something that actually made the brain behave differently despite the evidence of the senses, and some doctors held out hope that this new technique of determining brain wave patterns could help "tell you what manner of man you are" and have unique importance in psychoanalysis. (This theory was later discounted when it was reported that "similar patterns were produced by an eminent scientist and an English water beetle.") The breathless style of the *New York American* captured the thrill of discovery generated by Loomis' "newest diagnostic aid":

More than fifty years ago, Joseph Breuer of Vienna cured a 21-year-old girl, Anne, of a paralysis of the right arm by suggestion made after she was in a hypnotic trance. . . . Later on the famous Dr. Sigmund Freud of Vienna proved amply how unconscious wishes or desires brought about actual physical maladies. He dispensed with hypnotism altogether, and started his "psychoanalysis" method. Perhaps some day scientists like Loomis and his colleagues, Prof. H. Davis of Harvard and others, will employ their brain potential methods to investigate not only hypnotism but also psychoanalysis. What actually happens to brain activity rhythms under psychoanalysis? What is this mysterious thing called unconscious mentality? How is it released or inhibited? Will the new electrical instruments unlock these tantalizing mysteries?

Similar potential or action currents are recorded from nerves. Thus a most remarkable method is now available to probe into the mysteries of living tissues, especially nervous and brain tissues. Science has a new way of diagnosing brain tumors and other diseases, and eventually perhaps mental diseases will be thus investigated. . . .

Throughout this period, Loomis kept up his active exploration of a wide range of fields. In addition to electroencephalography, he was absorbed in measuring very small increments of time and was still playing around with his perfect clocks. He was also fascinated by the new field of high-energy physics and had even tried building a particle accelerator, known as a "cyclotron." Loomis would attack, with the same boundless enthusiasm he bestowed on the most important projects at the laboratory, innumerable other problems that caught his fancy, whether they were apparently frivolous or on the very fringes of science. He was interested in practically everything, and if he found a problem, at one time or another he probably pursued it, if only to try his hand at making some sort of headway.

"It was always about the next thing, the adventure," observed Caryl Haskins, who first visited the laboratory in Tuxedo Park in the late 1930s. "He never had *one* idea. He always had dozens of ideas. There were a lot of people working on things Alfred was directing and suggesting. I thought he was quite remarkable—a unique figure."

In 1937, Loomis embarked on one of his more curious extrascientific ventures when he collaborated with the innovative Swiss architect William Lescaze on the design of a state-of-the-art modern house. It would be the perfect marriage of form and function—embodying Loomis' ideal of a compact, climate-controlled, self-sufficient existence. Lescaze, who had arrived in New York in 1920, had made a name for himself designing forward-looking buildings that were influenced by the leading modern European architects, including Le Corbusier, Ludwig Mies van der Rohe, Walter Gropius, and J. J. P. Oud. He was most famous for the Philadelphia Saving Fund Society Building he designed with George Howe in 1932. The sleek, monolithic thirty-three-story tower, which housed one of the oldest and most conservative banks in the country, created a huge controversy, with critics divided as to whether it was an eyesore or heralded a brilliant new era of architecture.

Either way, the result was arresting, and by the mid-1930s, Lescaze had emerged as the leader of the new international style. A sunny stucco box with a steel frame he designed in a wooded enclave in New Hartford, Connecticut, had caused a sensation, particularly since his client, a young Vanderbilt heir named Frederick Vanderbilt Field, had demanded that the house represent a complete break from the dark, ornamental mansions of his childhood. Lescaze's structure was hailed as a modern classic, and architecture magazines of the day devoted pages to it. In 1934, Lescaze designed a Manhattan town house and office for himself and his family and was the first to incorporate sheer glass block walls and a built-in air-conditioning system in American residential architecture. The stark white house with its horizontal strip windows stood out from the dowdy row of brownstones like a bright beacon of the future, and with its spare interior design and custom furniture and cabinetry, it was as brilliantly functional inside as the cockpit of a plane.

At the time, Hobart, who needed a larger residence for his growing family, hired Lescaze to build a modernist stucco house in Tuxedo Park. During the architect's frequent trips to the area, he struck up a friendship with Loomis, and the two men found they had much in common. Lescaze, like Loomis, was a fierce perfectionist and was always redesigning common objects—everything from burglar alarms to pool tables—when he found their original form unsatisfactory or, as was more often the case, unsightly. Rather than let it spoil his plans, he redesigned the offending object. Once, when drawing up plans for a new school, he included a sketch of a "dustless blackboard eraser." Inevitably, the designs he worked out were simpler and more practical. Enthralled by their discussions of industrial design and new advances in technology, Loomis decided to collaborate with Lescaze on the design of a futuristic glass-and-steel house behind the laboratory. According to Lescaze's notes, the fundamental scheme of the house was dictated by Loomis' desire "to experiment with a novel system of heating and air-conditioning" and to be able to conduct these tests over a longer period than laboratory research allowed and "in ordinary living conditions."

The Glass House was to be the antithesis of the medieval Tower House, with its gloomy wood-paneled rooms, cathedral ceilings, and old-fashioned leaded windows. In typical Lescaze style, almost all of the structural components were made by machine, including the steel framing, cork flooring, and metal skylights. The single-story building's central section opened onto a large living room and conservatory, and from it sprouted a wing with two bedrooms and baths, and a second wing that consisted of a kitchen, utility rooms, air-conditioning room, terrace, and garage. Although the plan included two maids rooms, the Glass House was designed for a "servantless existence" and boasted the ultimate in modern conveniences, some of which Loomis designed himself. "It had built-in tubes for vacuuming, and the first dishwasher I think I had ever seen," recalled Evans. "I don't think Alfred ever wanted to see another one of his wife's meddling housemaids."

The most remarkable, and probably unique, feature of the structure was its double exterior walls and roof, so that in effect it consisted of one house built entirely within the shell of another house. The space between the double walls was approximately two feet, creating a corridor of air, or "shell space," that could be heated independently of the inner house. Since much of the house featured large glass panel windows and walls, the temperature of the shell space, if no heat was added, would be only slightly less than halfway between the outdoors and indoors. The object of this construction, according to Lescaze, was so Loomis could maintain a high temperature and humidity within the inner house without creating condensation on the glass. Loomis apparently wanted to try to re-create the balmy conditions of his home in Hilton Head, and according to Lescaze, one of the purposes of the experiment was "to investigate the effect of living in such an atmosphere during the winter season."

The Glass House was equipped with a special air-conditioning system, including an all-year unit for the inner house and a separate heating unit for the surrounding shell space. The complex duct system, which allowed return air and fresh air to be mixed and thermostatically controlled, and the oil-burning water heater and water cooler were worked out by Loomis and Lescaze and Leslie Hart, a consulting engineer. Owing to the "house within a house" construction, and heavy insulation used to deaden the sound of the mechanical equipment, the interior rooms were virtually soundproof. The worst Tuxedo rainstorms were barely audible within the building. Loomis boasted that the Glass House cost "only $125 a year to heat," a fraction of the sum he squandered annually to keep the drafty Tower House at a habitable temperature.

In his novel, Richards lampooned Lescaze's creation, which Loomis spent $125,000 to build, and another $25,000 to furnish, as a trumped-up "garden hot house." Apparently both architect and scientist forgot to allow for the fact that the miracle of modern air-conditioning could break down—as it often did during the summer months—turning the structure into a veritable oven. Because the double glass windows could not be opened, it took the rooms "days to cool off."

N O T long after the construction was completed in 1938, Loomis fell hopelessly in love with Hobart's twenty-nine-year-old wife, Manette. The Glass House, originally intended as guest quarters for visiting scientists, became their secret hideaway. Over time, it became Loomis' home away from home. Ironically, the house with translucent walls proved ideal for private rendezvous. Tucked away behind the laboratory on a secluded bluff overlooking Tuxedo Lake, and shielded on the other side by tall pines, it was protected from the prying eyes of neighbors. More than one of the lab's eminent guests was known to bring a mistress there for a "naughty weekend," according to Kisti-akowsky's second wife, Elaine. "It became quite the place for wild par-ties, and it was not uncommon for people to bring out their girlfriends and have quite a good time without their wives being any the wiser for it. They were all young, and quite good-looking, and they worked hard and played hard. And I gather they drank like fish."

Apart from visitors, Loomis allowed only trusted members of his lab-oratory staff access to the house and instructed his own wife that it was off limits both to her and to her legions of servants. "He left strict orders that no one was ever to go in there to tidy up, supposedly because of all the special equipment that was lying about," said Evans. "I don't know if Ellen knew what was going on or not, but she always *hated* that house."

Loomis went to great lengths to ensure he and Manette were not dis-covered, even developing a signaling system that he used to communi-cate with her from the windows of their respective homes. The Hobarts lived on the opposite side of Tuxedo Lake, and Lescaze had situated the house perfectly on the cliff so that its windows faced the water and of-fered a fine vista of the mansions on the other side, including the Tower House, rising from the top of the highest hill in the park. Loomis taught Manette how to manipulate a small mirror to catch the light and worked out a series of simple signals they used to alert each other at the appointed hour that the coast was clear. No matter how many times she heard the story, Loomis' granddaughter Jacqueline Quillen was always struck by the image of the two lovers secretly flashing messages to each other across the lake. "It was a very passionate love affair," she said, adding, "and despite all the trouble it caused, it remained that way to the end."

Manette was the daughter of R. W. (Billy) Seeldrayers, a prominent Belgian lawyer and sports promoter, who became head of the Belgian Olympic Committee. She grew up in a world of jocular athletes and, by her own account, learned at an early age "to enjoy male company far more than women's." Her father pushed her to excel at a wide variety of sports, and she received instruction in everything from tennis, golf, and field hockey to soccer and even a little cricket. She became an accom-plished tennis player and figure skater and briefly competed at the ama-teur level before giving it up to study music and art. Her family had lost most of their savings during the First World War, and her mother, ambi-tious for her only child to make a good marriage, tried to introduce her to "better society."

Manette met Katherine Grey Hobart in Brussels while the latter was on a European jaunt, and when the granddaughter of a distinguished American vice president invited her to return home with her, Manette's mother packed her bags. The Hobarts were exceedingly wealthy and divided their time between Carroll Hall, their elegant city residence in Paterson, New Jersey, and Ailsa Farms, the family's 250-acre country estate in Haledon. The Hobarts employed an army of servants, and the household staff alone included a cook, a kitchen maid, a parlor maid, a houseman, a butler, a laundress, an assistant laundress, two chauffeurs, and several chambermaids. They hosted "fancy dress" parties year-round at their stately forty-room mansion, and their table sparkled with Venetian glass and precious Fabergé Russian enamelware that had been designed for the czar. For twenty-two-year-old Manette, who had grown up in war-deprived Belgium and could still bitterly recall having a winter coat cut from the green felt cover of a billiard table, it must have seemed positively idyllic. In the space of a year, her betrothal to the Hobart's only son and heir was duly accomplished. They were married in Brussels in 1931 and divided their time between Ailsa Farms and Schenectady, before settling permanently in Tuxedo Park.

Never were two people more ill suited than the taciturn Hobart and his bright, athletic, puckish young bride. Garret Hobart was "pathologically shy," according to family members, and led a quiet, almost cloistered existence. He was quite content in his own little world, and the couple never entertained and had virtually no life outside the laboratory. While his neighbors considered him a bit queer but harmless, they steered clear of his "foreign" wife. Tuxedo Park was very provincial in those days, and anyone with an accent was seen as suspect. Manette's English was less than perfect, and she retained a thick Belgian accent that lent her a decidedly exotic air that the wives in the young smart set found most offputting. As a result, she had few friends and spent much of her time on her own. A talented artist, she spent her days working on her painting and sculptures, but it could not have been easy. "It was all very new to her, and she didn't really know a soul or how to get on," said Paulie Loomis. "I think the early years of her marriage must have been very lonely."

As her husband had no interest in sports, Manette took to playing tennis and golf with the young research scientists at Tower House, and more than a few became quite smitten with her, including Bill Richards. Very petite and slender, she had a superb figure that she displayed to full advantage. Although she was only passingly pretty, her emphatic sexuality made her captivating to the opposite sex. "Oh, she had a real way about her," recalled Evans, speaking with the authority of a southern belle who turned plenty of heads in her day. "She had wonderful legs, and always showed them off in little tennis skirts and golf shorts. She knew what she was doing. She was a real flirt."

Manette had a talent for making men fall in love with her, as her marriage to the unlikely Hobart attested, and she was not above using her sexuality to attract the fifty-year-old Loomis. "She absolutely seduced him," said Quillen. "I think it was about great sex, which would have been a scarce commodity in his first marriage. I think for Alfred it was an incredible, all-encompassing discovery. She gave him such enormous pleasure, and he absolutely adored her."

It is impossible to say exactly when the affair began. Both Manette and her husband were an integral part of the Loomis household and remained that way long after their relationship began. After Alfred made Garret Hobart his assistant, Ellen Loomis had taken his young wife under her wing and regarded her almost as a daughter. The two men worked together by day, dined together with their wives on a regular basis, and frequently took their families on holiday together. The Hobarts' first child, Garret Augustus Hobart IV, was born in 1935, followed by another boy, who was born in August 1937. Manette named her second son Alfred Loomis Hobart, after his beloved godfather.

According to Paulie Loomis, both Alfred and Manette were deeply unhappy for years before they became involved. "I know she was mad about him for a long time," she said. "Alfred was a wonderful-looking man, and very courtly and gentle. He could be hard to talk to unless he liked you. But once you got to know him, he was fascinating. He could explain the most complicated things and make them simple and understandable. He could unlock the secrets of the world, and it was magical. Manette was nobody's fool. Here she was married to poor Hobart, who was really an odd duck, and quite pathetic. She knew Alfred had no one, because his wife had taken to her sickbed long before that. And she knew he was the kind of man who just had to be with somebody. So she became his mistress, and she stayed married to Hobart. That was the cover-up, and I think it went on that way for a long time."

There is a striking black-and-white photograph of Manette and Loomis in a canoe that was taken in the summer of 1938 or 1939. She is happily reclining in the middle of the boat behind Loomis, who is paddling. The photo has been crudely cropped with scissors, cutting out the other oarsman, but in all likelihood it was Garret Hobart. The picture was taken at the Hobart family compound in Rangeley, Maine, the last time they were all on holiday together. "I am only guessing, but I don't think, at the time, my dad had a clue what was going on," said Al Hobart. Ellen Loomis' letters during this period reveal that she was lately "so hampered by illness" that she was not able to get out much or see friends, and it is possible she was unaware of the romance or simply chose to turn a blind eye to it. However, her condition became quite perilous again the following winter, which may have been her way of coping with the competition. As she wrote to Stimson in February 1939: "All my fever seems over now, and I know Alfred has given you the news. There is no cause for worry about me, as you always understood. . . ."

Garret Hobart never talked about the affair between his wife and revered mentor that eventually broke up his marriage. Only once, many years later, in a moment of frustration, did he betray a hint of the anger or bitterness he must have felt. "We were in Maine, and we were getting ready to go fishing, when he said, out of the blue, 'Alfred Loomis broke the tip off my fly rod,' " recalled Al Hobart, who was seven years old when his mother finally left his father to run off with Loomis in 1944. "That was it. Just that one outburst. But I caught the whiff then of a fairly strong resentment."

In his roman à clef, Richards, who was a good friend of Hobart's, caricatured Manette as the "brazen hussy" Leone Allison:

Her wide-set brown eyes and amiable expression were photogenic. The sun had bleached her hair until it was almost white and had turned her skin, most of which was visible, to a rich brown. She wore a rudimentary halter of robin's egg blue, tiny shorts of the same color, and rope-soled sandals. A wire-haired fox terrier with a handsome moustache and an aristocratic vacant expression trotted past her into the room. Every one turned as she halted in the doorway. . . .

Not only was her informality of dress "deeply shocking," her sexual frankness, for a woman of her day, was so surprising that it made grown men blush and left them "sputtering incoherently." She enjoyed playing "cat and mouse" games with various prey, but occasionally those she toyed with ended up falling in love with her, only to have their hearts broken. Throughout the novel, she boasts of having had affairs and admits to having an ill-considered fling with Bill Roberts, Richards' fictional alter ego, whom she "slept with . . . a couple of times." Roberts, she explained, had moved back to Boston and was unhappy and "drinking." But he could be very charming and persuasive, and she fell for him: "I was the only person he'd ever cared for, he said, and not having me was wrecking his life." After he had bedded her, however, it turned out he was not in love at all but had wanted only to add her to "his collection," and the two had a bitter parting of the ways.

In the novel, which Richards populated with cardboard cutouts of the famous scientists who frequented the laboratory, Leone Allison's husband, Charles Allison, the laboratory director (Garret Hobart) was portrayed as a weak-kneed, tremulous nerd who married a woman many times out of his league. As Leone (Manette) confesses in the book, her husband was aware of her infidelities, but there was nothing he could really do to stop her: "When I married Charlie I told him he wasn't the first, and he wasn't going to be the last." Nevertheless, she felt sorry for him and tried to protect him. "He thinks he just has to suffer if he doesn't like something. Why, even when he's making love, poor kid, he's sort of shy and all by himself." The only man who was truly her match, she admitted in a moment of candor, was her husband's boss— the wealthy owner of the laboratory. "Nobody else around here appeals to me . . . I've always been goofy about him, but he's happily married."

Even if Richards had lived long enough to insist that his novel was not intended to "represent persons living or dead," as he wrote in his author's note, his thumbnail sketch of Manette was entirely too vivid not to be instantly recognizable to the denizens of Tower House. By all accounts, it was dead on. "Oh, it was her all right," said Evans. "As soon as you read about her parading around in short shorts, you knew." When the novel was published in the spring of 1940, Loomis was appalled at the way Manette was depicted and outraged that Richards had dared suggest in print that there was anything between them. Beginning with the laboratory setting (using a private research facility housed in a mansion as the site of the murder and a vehicle for an in-depth look at the science of brain waves) to the catalog of familiar characters and painfully personal details, Loomis knew Richards had hardly invented a single element of his story. "Alfred hated that book," said Evans. "He absolutely hated it. He wished that it had never been published."

Exactly when Loomis became aware of the book is not clear, but

Richards' stunning suicide just weeks before its publication apparently cut short any legal action Loomis might have contemplated taking to quash the inflammatory novel. Although Richards' family worried that Loomis might still sue for libel, it seems unlikely, as it would surely have attracted further publicity, which was the last thing he wanted. Besides, Richards had disguised the Loomis Laboratory well enough, and nothing was ever written in the newspapers about the fictional story's surprising similarity to his Tuxedo Park establishment or the important brain wave research being done there. As far as Loomis was concerned, the best thing to do was to bury the book along with its author. He never spoke of either again. There was a rumor at the time, according to Richards' nephew Ted Conant, that Loomis bought up every copy of the novel available in New York bookstores, just to make sure that as few friends and acquaintances saw it as possible. But as the deeply chagrined Richards family also wished the book would disappear, no one would have stopped the powerful Wall Street financier from doing as he saw fit.

In all fairness, Richards' suicide must have been deeply shocking to Loomis and his family. He had been a close friend and colleague. He was among the very first of the young scientists Loomis had recruited to work with him at Tower House, and their association had lasted over fourteen years. He was still working for Loomis on a part-time basis at the time of his death, and they must have been in regular contact. Certainly, Richards had suffered bouts of depression, had occasionally drunk to excess, and had often been physically unwell, but none of those things had made him decidedly more peculiar than any of the others in Loomis' company.

Kistiakowsky, who was at Harvard by then, had been very close to Richards since their Princeton days and had known more about his "mental troubles" than anyone. Richards had confided to him in intimate detail about his tempestuous personal life. He had had a series of failed love affairs, including one with Christiana Morgan, the beautiful but volatile daughter of a Boston society family. Morgan had become a protégé of Carl Jung, and Richards, who was very taken with her, had followed her to Zurich and had even consulted Jung about his sexual problems, which he blamed on his repressive Puritan background. When Richards quit his teaching post at Princeton, he had told Kistiakowsky that his emotional state was worse and he was moving to New York to undergo intensive psychotherapy. (It is probable that Richards was manic-depressive: his father, the Harvard Nobel laureate, had suffered from myriad phobias and "nervous attacks" and had died at the age of sixty after being laid low by chronic respiratory problems and a prolonged depression; years later, Bill Richards' brother, Thayer, a prominent Virginia architect, would also commit suicide, lying down on the railroad tracks near his home.) On the last page of his novel, Richards has a character ask one of the doctors at the laboratory, "You have all heard the expression 'Genius is close to insanity.' Do you believe, as a psychiatrist, that this is essentially a representation of fact?"

After Richards' suicide in January 1940, Kistiakowsky could not duck the guilt he felt over the role he played in his friend's increasing dependency on booze. "He was an excellent conversationalist, well-versed in cultural and artistic matters, a gay companion in all the drinking parties we used to have," Kistiakowsky later recalled in his memoir. "Meanwhile he became an alcoholic. I fear that our joint drinking of bootleg alcohol that we used to doctor up into 'gin' was a critical stage on that road. . . ."

It is doubtful Loomis ever suffered any such misgivings about Richards' death. He had already moved on. The past was done with, and all that mattered was the future. By then, he had met his new protégé, Ernest Lawrence, and was impatient to see what they could accomplish together. Paulie Loomis always admired her father-in-law's relentless quest for scientific truths but could never completely ignore its ruthless quality. "Physicists are single-minded in the pursuit of what interests them," she observed. "As people go, they can be pretty cold."

WORLD events also conspired to distance Loomis from the lofty pursuits and low intrigues at Tower House. By the late 1930s, as the Nazi assault on Europe gained momentum, Loomis' scientific interests began to change. He once again became obsessed with Germany's artillery and machinery of war, just as he had in the years before World War I. He had also had disturbing reports of the staying power of Mussolini and Hitler from the physics grapevine and his many foreign guests, including Bohr and Fermi. Loomis made several trips to Europe, and in 1938 he traveled to Berlin and visited Bohr more than once in Copenhagen. He was very troubled by what he learned about the highly developed state of applied scientific research in Germany. From Kistiakowsky and others who were interested and knowledgeable about Germany's efforts to rearm herself, he heard unsettling things about advanced weaponry and the work German physicists were rumored to be doing in nuclear physics.

From his regular conversations with Stimson, who had returned to private life and his law practice, Loomis knew that Congress, reflecting the general sentiment of the country, was determined to stay out of the mess in Europe. He also knew that his cousin was of the opinion that the best way to avoid war was to remain alert and not abdicate responsibility, that isolation, in the modern world, was impossible. As Stimson had argued in a letter to *The New York Times* on October 11, 1935, and reiterated in a radio address on October 24, "The real problem is to decide what methods of action will best keep us out of war" at a time when "civilized life has suddenly become extremely complex and extremely fragile," when "the world had suddenly become interconnected and interdependent." While Stimson had never publicly come out against the administration's position on neutrality, in private he maintained that the existing policies at home and abroad, if continued for long, would surely lead to catastrophe.

In 1936, exactly one week after Roosevelt had signed the second Neutrality Act, Hitler marched into the Rhineland. It was a flagrant violation of the Treaty of Versailles, which had established the Rhineland as a buffer zone between Germany and France. In the ensuing months, civil war broke out in Spain. After Japan renewed its aggression against China, Roosevelt delivered his famous "quarantine speech" in October 1937, a first cautious call for the reining in of "lawless aggressors." Stimson, roused by the president's preference for talk over action, wrote a letter to *The New York Times* calling for leadership and faulting the administration's "ostrich-like isolationism" and "erroneous form of neutrality legislation [which] has threatened to bring upon us the very dangers of war which we are now seeking to avoid."

Loomis had overheard enough of his colleagues talking to know that most Americans felt they had done their part to save democracy in the First World War, and now it was Europe's responsibility to solve its own problems. Einstein made headlines when he immigrated to America in 1933 to escape the Nazi tyranny, and since then many Jewish physicists who worried that they might soon be forced to leave their teaching positions had fled the war-torn continent. In December 1938, Enrico Fermi, whose wife was Jewish, left Rome to go to Stockholm to collect his Nobel Prize and never looked back, traveling to New York, where he took up a position at Columbia University. But for scientists in the

United States, without close friends or relatives abroad, foreign problems were a matter for politicians and policy experts—and they generally shared the cheerful view that if Germany wanted to let its brightest minds leave, it was their loss and America's gain. Loomis was not as sanguine. He had talked to his old friend Compton and had seen Conant at Harvard, and both were very pessimistic over the fate of scientists in Germany, Austria, and Italy. He could not help but share his cousin's view that the country had crept into a hole and was trying to forget the world.

Loomis believed that if Europe was going to fall apart, it was far better to be vigilant. It was a lesson he had learned back in his days at Aberdeen, testing Edison's theories on the best ways to cope with the deadly U-boats. After the shock of the sinking of the *Lusitania* by a German submarine in 1915, the famous old inventor had exhorted the country in *The New York Times* that Americans were "as clever at mechanics . . . as any people in the world" and could defeat any "engine of destruction." Edison had advocated preparedness without provocation, and to Loomis, it seemed as wise a course in the present as it had been then.

When Hitler rolled into Austria in 1938, and then decimated Czechoslovakia, Loomis made note of the tank models, the destructiveness of the field artillery, and the brutality of the bombings. His exposure to army procedure at Aberdeen, the antiquated cannons and hidebound bureaucracy, had left him convinced that the military could not be counted on to develop and build a stockpile of modern weapons for defense. At the start of the last war, Edison had recommended that the government create "a great research laboratory" whose purpose would be to develop new weaponry, so that if war came, the country could "take advantage of the knowledge gained through this research work and quickly manufacture in large quantities the very latest and most efficient instruments of war." In the months to come, these accumulating influences would move Loomis to adapt Edison's ideas to his own laboratory.

A call from Compton in early 1939 helped crystallize his plans. Compton had correctly sensed that Loomis was at loose ends and was casting about for a new direction for his research. He had told Compton that his work on brain waves with Hallowell Davis, while far from complete, needed to be carried on under the auspices of a hospital, and to that end he had donated most of his equipment to the Harvard Medical

School. Compton suggested that, given the portentous events in Europe, it might be useful if Loomis looked into the present state of microwave radio technology, or radar (though the latter term had not yet been coined). Loomis was intrigued by the idea and began exploring the subject on his own.

Compton, of course, had his own reasons for wanting to involve Loomis. MIT had for some time been doing exploratory work in the study of microwaves, but their program needed additional financial support if it was to continue, and both he and Vannevar Bush were hoping that Loomis would step up and provide the funds for a joint research project. Earlier that spring, he had arranged for Edward Bowles, the MIT radio specialist who was largely responsible for the university's research program, along with several of his top investigators, to meet with Loomis at Tuxedo Park. Bowles was bright but temperamental, and had managed to alienate a number of colleagues over the years, including Bush. But he was also a keen enough promoter to have kept MIT's blind-landing radar operation going from grant to grant, including one from the Sperry Gyroscope Company, and was employing some fifteen MIT investigators, all working on ultrahigh-frequency microwave projects. Bowles and Compton together convinced Loomis that the field showed great promise, and he signed on, offering his laboratory as a research facility.

In their correspondence that winter and spring, Loomis and Bush discussed "the matter of distance finding by radio." In early February, Loomis wrote to Bush that "Mrs. Loomis has been quite ill with undulant fever," necessitating a trip to Honolulu, but that on his return he planned to stop over in California and "am wondering what scientific laboratories, etc. you would suggest that I visit." After Loomis' Hawaiian vacation was canceled because his wife was too sick to travel, he invited Bush to meet with him in Tuxedo. Afterward, Bush wrote Loomis:

> I will take up with Bowles the possibility of developing some simple equipment for approximate distance finding at the same time that precise equipment is being developed . . . thus that we may be able to start the program a little more rapidly than would otherwise be possible.

A few weeks later, Bush followed up with a long letter, addressed "Dear Loomis," explaining in great technical detail his idea of "how we might make a plane detector," including "a fairly simple computer, which would control the gun directly." Bush, who was mechanically inclined, and while at MIT had invented his famed "differential analyzer"—an early computer that did intricate mathematical calculations—clearly felt he and Loomis spoke the same language, noting at one point, "The trigonometry involved is not bad." At the end of the letter, he asked Loomis to "keep it confidential," adding, "This is all very sketchy, but it may have a lead in it somewhere."

That summer, Loomis joined Compton at a conference on ultra-short-wave radio problems held at MIT. The symposium was attended by the representatives of all the major companies doing research in the field, but they were so excessively guarded and tight-lipped about the details of their patents that Compton dismissed the formal presentations as "pathetic and amusing." However, that evening at dinner, after a considerable amount of teasing, and no doubt drinking, they were gradually induced to tell their stories and reveal some of the real progress that had been made.

In the 1930s, microwave research was heavily cloaked in secrecy and was simultaneously being developed under wraps in military and industrial laboratories in America, England, France, and Germany. The basic principle, that radio waves had optical properties and could "reflect" solid objects, had been demonstrated in 1888 by the German scientist Heinrich Hertz. A working device for the detection of ships, based on his experiments, was tested in the early 1900s. But little was done to exploit the discovery, even though as far back as 1922, Guglielmo Marconi had urged the development of short radio waves for the detection of obstacles in the fog or darkness. It was not until the 1930s, when airplanes came of age as a military weapon—a threat made terrifyingly real by the damage inflicted by German and Italian bombers on Spain between 1936 and 1938—that the technology of radar finally began to be developed in earnest.

Most of the countries exploring radar concentrated their early efforts on "the beat method," or the Doppler method, which used ordinary continuous radio waves and required at least two widely separated and bulky stations, one for transmitting and one for receiving. Airplanes that penetrated between the transmitter and receiver were detected by the Doppler beat between the direct signal (from the transmitter to the receiver) and the signal scattered by the target (which traveled a longer route from the transmitter to the plane and then to the receiver). Unfortunately, the equipment was fairly limited in its effectiveness. The sharpness of the system's vision—its ability to distinguish separately the echoes from two targets close together and at the same distance from the radar—depended on the sharpness of the radar beam. For a given antenna, the beam width was proportional to the wavelength and would become sharper as the wavelength decreased. Loomis realized that if sharp radar beams were ever to be produced by an antenna not too large to carry in an airplane, they would have to develop a generator of much shorter wavelengths than was then known. It was speculative, to be sure, but the unexplored microwave spectrum promised not only to allow radar sets to become much smaller and more portable, but also to prove better at locating low-flying aircraft and to be able to distinguish targets with far greater accuracy.

Loomis, operating in a manner that Compton described as "typical of him," spent the next few months quickly mastering the new subject and "worked with his little permanent staff at Tuxedo on the fundamentals of microwave until he felt capable of inviting collaboration." Late that summer, Bowles and a group of MIT physicists arrived at the Tower House and began an in-depth study of the propagation of radio waves. The main feature would be a study of ultrahigh-frequency propagation, to be conducted by J. A. Stratton and Donald Kerr, veterans of MIT's blind-landing research program, "to determine the practical range that we can expect to obtain with 50 cm waves, which we now have facilities to generate." As they progressed, they would apply their techniques to shorter and shorter wavelengths.

Bowles was not at all keen on the idea of working for the retired Wall Street banker, with his perfectly pressed white suits and "ideal living and laboratory quarters." The "Tuxedo Park situation," as he called it, was more "complex" than he had first reckoned, and he privately suspected the financier had invited them only because it put a small company of MIT scientists, himself included, at Loomis' disposal, "pretty much to follow his bidding." Much to his dismay, "Loomis himself was a gadgeteer and pretty much called the shots." But Bowles was too ambitious to rock the boat and tried his best to humor his new boss. "It was simply that I knew he was a close friend of Karl Compton's, and, no doubt, this summer's activities had his benediction. What I got out of it was some knowledge of Loomis, his technical interests, and his manner of operation, [of] which I was later to learn much more."

Loomis, on the other hand, could not have been happier. Pleased to be of service, and thrilled by the challenge of perfecting this critical new technology, he dropped all of his other experiments to concentrate on the microwave project. In the process, he drastically rewrote the charter of the Tower House. Once a bastion of pure science, Loomis' laboratory, tucked away in the lush hills of Tuxedo Park, was on its way to becoming a private research center devoted to the development of secret war-related technology—the radar systems used to detect airplanes.

Chapter 7

THE BIG MACHINE

Remember how Aston says, in his book, something about
"when man has unleashed the energy of the nucleus, the result
will be published to the universe as a new star"?

—WR, from "The Uranium Bomb"

ERNEST LAWRENCE first ventured out to Tuxedo Park
for an extended weekend in 1936. "Just to meet and talk about things,"
Loomis recalled, and "to see the lab." Loomis had not invited him personally, so the gregarious Berkeley physicist must have tagged along as
somebody's guest. It may have been for one of the annual meetings he
held at Tower House, when dozens of scientists came from all over the
country. There was nothing unusual about his wanting to see the place,
"because almost every famous scientist had been out there as a routine
thing."

Of course, Loomis knew Lawrence by reputation. He had earned international acclaim at the age of thirty for his invention of the cyclotron, the most powerful machine for smashing atoms ever built, and
was reportedly making formidable use of the device. Known for his brilliant inventive mind and a boyish enthusiasm that was almost contagious, he was widely regarded as one of the most promising young
physicists around. Still, nothing prepared Loomis for the jovial and

easygoing fellow who ambled up the drive and introduced himself. After meeting Lawrence, Manette declared the tall, blond, blue-eyed Swede, who happened to be a top-notch tennis player, to be one of the most charming men she had ever met: "He was completely opposite of what you expect a scientist to be—he was just a handsome big fellow, full of loving and full of fun, and very easy to be with. You were friendly in five minutes."

At thirty-five, Lawrence had a winning manner, determination, and zeal that had already made him something of a legend in nuclear physics, which was the new fad sweeping the physical sciences. A native of South Dakota and a graduate of its state university, he had followed his mentor, Merle Tuve, a fellow South Dakotan who was also of Scandinavian descent, to the University of Minnesota to work on his master's degree. As a National Research Fellow at Yale, Lawrence had published a series of papers on "ionizing potentials"—demonstrating that an electron will ionize an atom—that had quickly brought him attention as a exceptional experimental talent. In 1928, Berkeley, which was in the midst of an ambitious drive to expand its Physics Department, managed to lure him out west (and away from Yale!) with the promise of fast promotion and ample funds for equipment. The following year, he conceived of the basic idea of building the cyclotron after reading a paper by an obscure German scientist on the behavior of ions in a magnetic field.

In 1931, Lawrence and his co-workers succeeded in building the first cyclotron, using a tank six inches across and a small electromagnet whose poles faced each other vertically across the gap. In the gap was placed a shallow cylindrical tank, pumped out to a high vacuum so that the particles inside could move freely without interference from air molecules. Lawrence fed deuterons (heavy hydrogen nuclei) as atomic projectiles in at the center and kicked them around at high speeds using a radio frequency oscillator. He then graduated to a bigger setup, using a huge eighty-five-ton magnet and a vacuum tank eight inches across, which allowed him to accelerate the deuterons at very high speeds and direct them against any target. His work developing powerful beams of particles had already earned high praise from none other than Bohr himself, "the dean of quantum theorists," who would make two trips from Copenhagen to California in the 1930s to check up on the young Berkeley physicist.

Right from the start, Loomis was "very impressed" with Lawrence, whose work in the field of high-energy physics held special interest for him. After all, the rapid pace of development of high-energy physics at Berkeley was due largely to the parallel development and expansion of an industry Loomis had been involved in since its infancy—hydroelectric power. Southern California Edison was the world's largest producer of hydroelectric power, and throughout the booming twenties the utility had given millions of dollars for physics research programs at Berkeley and Cal Tech to find ways to improve the technique of high-voltage transmission. Loomis was something of an expert on the need to transmit power economically, having worked that equation on Wall Street for the better part of a decade, and over the years he had kept a close eye on the advances in high-voltage technology.

Physicists had wrestled for years with the problem of achieving high voltages for their scientific investigations. After a scientific conference in 1928, Loomis had talked with Sir Ernest Rutherford, the revered director of Cambridge University's Cavendish Laboratory, about the difficulty of producing and controlling big voltages on the order of what nature packed in a bolt of lightning. Lord Rutherford had done the pioneering studies of radioactivity and, with others, had found three kinds—alpha, beta (streams of electrically charged particles), and gamma (as in X rays)—which had led to his formulation of the nuclear model of the atom. The gruff, burly British Nobel laureate was convinced that the creation of machines operating at the highest possible voltage was "a matter of pressing importance" and had taken the lead in promoting the development of million-volt accelerators. "Rutherford was an old man, and very abrupt in conversation," Loomis recalled. "I [had] just met him and we were talking, and he suddenly burst out and said, 'You damned American millionaires. Why can't you give me a million volts, and I will split the atom.' " Loomis, who shared the great man's frustration, could only reply warmly, "Well, we don't know how to make a million volts that can be useful to you. We know how to make sparks jump, but we don't know how it's going to be useful."

Loomis' interest in high voltages prompted him to try his own cyclotron experiments. At one point, he and his colleagues at Tower House "broke down a quarter of a million machine," which struggled to produce 250,000 electron volts, just "to see what we could do." He had no trouble laying his hands on one, as he was a member of the MIT Corporation and was quite involved with the high-voltage machine the school had developed. According to Vannevar Bush, who was then vice president of MIT, Loomis, who was "a rather red-hot individual," once burst into his office and told him he had heard that their cyclotron, which had been installed in a Boston hospital, was always broken down and could not be fixed, and that "it was a terrible thing MIT couldn't build a good machine." Bush, who suspected the mischievous rumors had been spread by the General Electric Company, which manufactured high-voltage devices and was none too happy that MIT was invading their turf, had had to march Loomis down to the hospital himself to prove to him that their machine "operated nearly perfectly." So when Loomis later heard that Lawrence had succeeded in building a big cyclotron and "had gotten a million usable volts out of a little seven-inch disc," he understood immediately "just what [Lawrence] was working for and why he was working for it." It wasn't new to him, because at the very same time, he had been working "on a parallel track."

It was with that background that Loomis and Lawrence first talked in Tuxedo Park and, according to Loomis, immediately "hit it off." He admired the younger physicist's daring and inspired resourcefulness. Loomis asked his advice about an experiment he was working on, and Lawrence spent the day with him at Tower House and followed up with an invitation to visit his laboratory at Berkeley. After that, whenever Lawrence came east, he stayed with Loomis at his New York penthouse or came to Tuxedo for the weekend. During the mid-1930s, Lawrence was making special experiments on the thirty-inch to get test data and planning his next cyclotron. His ideas on just how to build the device were always shifting, according to Loomis, and every time he came to visit he would announce, "This is a better way than we talked about last month." Lawrence's enthusiasm was catching, and before long, Loomis was caught up in his plans to build another massive cyclotron. Both men were bonded by their innate optimism, a quintessentially American belief in technology that had as its rallying cry the macho credo of cyclotroneering—the bigger the machine the better. They shared an absolute faith in scientific progress. And for them, the pace of change could not be fast enough.

What gave urgency to their work was the pressing importance of the science itself, the novelty of nuclear transformation. Although the situation in Europe appeared more precarious every day, neither Lawrence nor Loomis had begun to think about the cyclotron in terms of its potential as a formidable atomic weapon. They were committed to pure research—pure curiosity. "We were obligated to science to exploit it, and every time we'd make it bigger, we'd find new facts," recalled Loomis. "Obviously, if the seven-inch worked and the thirty-inch worked, the next step, the quicker, the better, would be to go up higher."

The two men became best of friends in a single bound. It was a meeting of minds that Loomis once said was as simple and completely symbiotic as picking up a conversation that had no beginning, middle, or end: "Ever since we first knew each other, there was a continuity as definite as if we'd lived in the same building all the time. There were gaps in time that weren't any bigger than if I should go upstairs, change my suit, and come down. We would go right on where we had left off. And his dream of bigger and better was there from the very day he built that thing [the cyclotron]."

As Alvarez wrote, the relationship that quickly developed between Loomis and Lawrence had "all the earmarks of a perfect marriage":

> They were completely compatible in every sense of the word, and their backgrounds and talents complemented each almost exactly. . . . Lawrence had developed a new way of doing what came to be called "big science," and that development stemmed from his ebullient nature plus his scientific insight and his charisma; he was more the natural leader than any man I've met. These characteristics attracted Loomis to him, and Loomis in turn introduced Lawrence to worlds he had never known before, and found equally fascinating. Anyone who was in their company . . . would have thought that they were lifelong intimate friends with all manner of shared experiences going back to childhood.

Lawrence's star was then already on the rise, but it was not yet universal, and his allegiance with the wealthy and influential Loomis speeded him on his way. Each stimulated and built up the other, so that everyone who was drawn into their nexus felt a charge of excitement—the thrilling sense that anything was possible. What they wanted was to build a tremendous cyclotron. That it would also cost a tremendous amount never once gave them pause, and their boundless confidence not only sustained them, it spurred them on to win the richest prize in physics—the breathtaking $1.15 million grant from the Rockefeller Foundation.

In the 1930s, raising large sums for scientific research was a daunting task, and Lawrence had to devote an inordinate amount of his time to scrounging money and materials. Nearly all scientific research was privately supported, and during the Depression, there was limited public sympathy toward underwriting the expense of scientific knowledge. The technological advances that for so long fueled the industrial machine had manifestly failed, and the country felt not only betrayed by science, but deeply ambivalent about its impact on their lives. Mechanized factories were blamed for throwing tens of thousands of assembly-line workers out of their jobs, and politicians railed against the "grave maladjustments" the rapid pace of technological progress was wreaking on society.

As a result, research funds were scarce, and the competition was stiff. Lawrence's readiness to share his cyclotron technique with other laboratories had come back to haunt him in the form of many imitators: there were eleven cyclotrons in various stages of operation in America and roughly the same number in Europe and Japan, and many of these projects were directed or staffed by research fellows whom he had trained at Berkeley. Most of Lawrence's grants had come from universities and hospitals interested in developing cyclotrons for biomedical purposes. He had discovered that cyclotrons produced "penetrating radiations" that had shown promise in the treatment of certain types of leukemia, and if, as experiments seemed to show, neutron bombardments had a greater effect on tumors than X rays did, then Lawrence believed they had a new weapon to fight cancer. Lawrence's brother, John, a doctor, had collaborated in studies using direct radiation from the cyclotron for cancer treatment. It was with that pitch that Berkeley president Robert Sproul had appealed to William Crocker, a retired banker and major benefactor of the medical school, who was finally persuaded to fork over another $75,000 so that Lawrence could run his own program on campus, known as the Crocker Radiation Laboratory, or Rad Lab for short.

Berkeley had worked desperately to keep Lawrence, who was being courted by Harvard's president, Jim Conant. To steal Lawrence away, Harvard had promised to pay him a salary of more than $10,000, to create positions for Berkeley's other star physicists, Robert Oppenheimer and Edwin McMillan, and to provide ample funds for a laboratory that would be, in Conant's words, "second to none," including a very big cyclotron, cloud chamber (an expansion chamber that makes the path of a charged particle visible for further study), and other auxiliary equipment. Conant had even approached Loomis about becoming a trustee in hopes that currying favor with the wealthy patron would help their cause. Instead, the overtures from such a prestigious Ivy League suitor prodded Berkeley to dig deep into its pockets to provide Lawrence with "his heart's desire"—a new, greatly expanded laboratory and an even larger cyclotron. "I shall always be grateful to you for the honor of your confidence," Lawrence wrote to Conant, declining his offer. "It opened up possibilities for work here that a month ago I thought were quite out of the question."

In October 1937, Lawrence's biomedical research won the Comstock Prize of the National Academy of Sciences, regarded as the highest scientific honor in the country. The "Cyclotron Man" made the cover of *Time* magazine and was hailed in the headline for his godlike powers: "He creates and destroys." Inevitably, his extraordinary success, combined with his extreme youth, bred jealousy and led to nasty gossip that Lawrence was just using cancer to unlock foundation coffers. Loomis, who had always looked for practical applications for all of his research, dismissed the criticism as ridiculous. "[Ernest] was a great optimist that somehow or other this would work for medicine. It was obvious from the very beginning, when he was building [radioactive] isotopes, that it opened up matters for making medical measurements as well as chemical and physical measurements."

Even before the sixty-inch design—dubbed the "Crocker Cracker"—was completed, Lawrence was thinking of one "ten times greater," a truly huge cyclotron for nuclear physics or, as Isidor Isaac Rabi of Columbia called it, "the beam to end all beams." For Lawrence, money, not technology, was the chief obstacle. In a radio broadcast, he announced he was considering constructing a cyclotron "to weigh 2,000 tons and to produce 100 million-volt particles. . . . It would require more than half a million dollars." With the active encouragement of Loomis and other big-thinking admirers, it would increase steadily in size and cost over the next year. "He was building a cyclotron as big as money would permit him," said Loomis, adding that "we got up to 210 inches" before it was finally cut back to 184 inches. "The idea would go up and up. He did very courageous things. Most people would not want to make such a big calculation, but he was so confident."

At the time Lawrence was sketching his massive cyclotron, one that would make the Brits' Birmingham model look like a "toy," he was going directly against the accepted opinion of experienced cyclotroneers. At Cornell, Hans Bethe, along with his student Morris E. Rose, had con-

cluded that the thirty-seven-inch machine was optimum, and their calculations showed that relativity limited the maximum energies obtain-
able in a cyclotron. Anything bigger was a waste of time and money. In
some quarters, it was even hinted that the brash Swede was too eager for
fame and was overreaching. But Lawrence's great strength was that he
never bowed to accepted theory. As he once told a colleague, "It never
does much good to find out why you can't; put your effort into what you
can do." Lawrence persevered and coolly answered Bethe's criticism:
"We have learned from repeated experience that there are many ways of
skinning a cat."

Loomis had much in common with Lawrence in this respect, as he
too had little time for the superiority of theoretical physicists over the
experimentalists. "It was only, 'Aren't you trying to get too much
money? Aren't you trying to make too big a step?' " said Loomis. "And
that was easily answered by the people who knew about it. The real
question was whether you could get the vacuum tubes [big enough] in
those days, to step it up. Ernest's answer was, 'Do it anyhow, whether we
can get them or not, because we just have to buy 100 of these tubes and
put them together in a bank.' "

On September 1, 1939, Germany invaded Poland. Already allied
with Mussolini's Italy, Germany had that August signed a nonaggres-
sion pact with the Soviet Union, clearing the way for her march
through Poland. Two days later, Britain and France declared war on
Germany and Italy. The news shocked everyone at the laboratory, particularly Lawrence, who was inclined to block the troubled world from
his mind in order to focus on his work. There was great concern all
around, because Lawrence's brother was still in England, where he had
gone to give a talk on the Rad Lab's work on radiation treatment to the
British Association for the Advancement of Science, and was due to
sail home on the *Athenia*.

That evening, the radio carried a report that a German submarine
had torpedoed the *Athenia*, and it was sinking off Scotland. Sick with
worry, Lawrence sat glued with his ear to the radio for six hours until the
news came that all of the Americans on board had been rescued by a
British destroyer. But they could not be sure the reports were accurate,
and their doubts were made worse by conflicting reports that some five
hundred passengers and crew had been lost. Two more tense days followed, then Lawrence finally received word from an *Oakland Tribune* reporter that John was safe. Lawrence immediately made some calls to

Washington and cabled his brother in Glasgow that he had secured him a berth on the first American ship sailing from Britain. For the Berkeley physicist, there was no longer any escaping the reality of the war in Europe.

Lawrence wrote expressing his gratitude to the British physicist John Cockcroft, his friendly rival at Cavendish Laboratory, who had looked after his brother in London, and had forwarded a recent letter from him along with his own kind note. Despite the widening war, Lawrence still shared the widespread view that Europe would resolve its problems without involving the United States and that he should continue devoting himself to purely scientific pursuits and experimentation:

> We have all been through a harrowing experience. I could not bring myself to believe that hostilities actually would break out, and the attack on Poland came to me as a great shock, as doubtless it likewise did to millions of people elsewhere. It is all a very sad business, and the best we can hope for now is that it will not go on for many years. As you doubtless know, the feeling over here is practically unanimously on the side of the allies, and there is complete confidence that they will win in the long run. Let us hope that it will not really take a long time. . . .

LOOMIS arrived in San Francisco in early November 1939 and spent most of the winter on the West Coast, dividing his time between Lawrence's cyclotron and his work for Compton on radar. He and Ellen stayed at the Mark Hopkins, and after she returned east, Loomis moved into the Claremont Hotel in Oakland. He organized a small laboratory on Berkeley's campus to carry out a microwave experiment that Lawrence helped him design and even brought out some men from Tuxedo Park to work there.

The main purpose was to bring himself up to speed on the microwave radar system based on the klystron tube, which had been invented by an instructor at the Stanford Physics Department named William Hansen, with help from a former roommate, Russell Varian, and his brother Sigurd, who was a commercial pilot. Hansen was also a mathematical genius who had entered Stanford at fourteen, where his groundbreaking work on microwaves soon brought him attention. In the summer of 1937, the Varian brothers had worked with Hansen on the design of a microwave device for navigating and detecting planes. To get the wavelengths in the range they wanted, and to shrink radar equipment from its present unwieldy size, they had to overcome the frequency limit of commercial oscillator tubes. In a stroke of brilliance, Hanson had tried setting up oscillations in a cavity, and it worked. His cavity resonator, or "rhumba-tron," as he called it, became the basis for their first crude detection device, made of cardboard coated with copper foil, operating in the thirteen-centimeter range.

The Sperry Gyroscope Company sensed he was on to something and offered to sponsor a research program in aircraft detection, which led to Russell Varian's refinement of the idea and the design of the klystron tube, named for the Greek verb *klyzein*, the breaking of waves on the shore. They went on to develop a detection system using a ten-centimeter klystron, though with very little wave power, and a range of possibly three to four miles. They did early experiments with an antenna reflector consisting of a sixteen-foot parabolic cylinder mounted on a Sperry sixty-inch searchlight frame, which provided a very narrow, sharper beam. The device detected moving objects by the Doppler frequency of the moving object. As the Sperry klystron program expanded, it was divided into two projects: a small group, working under Hansen out of the Stanford Physics Department, was carrying out experiments on the airplane detecting device, while a much larger team, which had assembled at the spacious plant Sperry had built in nearby San Carlos, was trying to design a satisfactory blind landing system based on the klystron. Both groups were also doing research to try to improve the tubes. True to form, Loomis was Sperry's first customer. He had shown up at their door, checkbook in hand, just as he had years before to purchase one of the first Shortt clocks.

When he was not working on radar, Loomis spent every spare minute in the Rad Lab boning up on cyclotron engineering. After listening to all Lawrence's plans for an enormous 100–200-million-volt cyclotron, he declared himself convinced and threw his considerable support behind the giant machine. He also proved helpful in a variety of ways, in one instance using his industry contacts to help Donald Cooksey, assistant director of the Rad Lab and Lawrence's right-hand man, get copper and steel for a cyclotron under construction in Calcutta. Even more appreciated were the large dinners he regularly hosted at Trader Vic's and Di Biasi's attended by Lawrence and his wife, Molly, Cooksey, Alvarez, and the rest of "the gang" at the Rad Lab.

"Mr. Loomis would come out here to Berkeley for several months at a time and work right in the lab with Lawrence," recalled a senior Rad Lab colleague. "He knew his physics and was capable of working. You'd never think he was something special, except that he always arrived in a big seven-passenger limousine. His chauffeur just sat in the car the whole day, waiting for him." Nobody knew how much money he gave Lawrence directly for his cyclotron project. Lawrence once confided that Loomis was "one of the ten richest men in the country, yet nobody knew this." Everyone at the lab was aware Loomis had established a fund marked "Ernest O. Lawrence, Personal," as the checks were signed. "He made Lawrence his protégé and played angel to him all through the years. He gave Lawrence advice about investments, and I imagine Lawrence made quite a bit out of them."

Loomis kept a desk in the building where Lawrence worked, and the thirty-seven-inch machine was running right downstairs. The assault on the unknown territory of the nucleus had begun, and he could not stay away. "I just got caught up in it then," he recalled. "It was all the excitement. The soldiers had been getting across the bridge into a new field." Alvarez, who was one of Lawrence's protégés at the lab, was not surprised to see the "millionaire physicist" on the premises, as he had heard much about his activities from Berkeley colleagues who had been to Tower House. A tall, ruddy blond—he favored his mother's family, which was Irish, rather than his father's Spanish heritage—Alvarez was assigned to show Loomis around. But after taking some time to introduce their esteemed guest "to more nuclear physics than he had known before," Alvarez was more than a little taken aback by Loomis' response:

I [had] mentioned in passing that because of the war in Europe the price of copper had risen to almost twice that of aluminum, volume for volume. Since aluminum has only 60 percent more specific resistivity than copper, I suggested that aluminum might now be the preferred metal for the magnet windings of the 184-inch cyclotron. It seemed obvious to me from elementary scaling laws that an aluminum coil would be larger but would cost less. I had completely forgotten about the suggestion when, a few days later, Alfred showed me a long set of calculations based on several altered designs of the 184-inch cyclotron that proved my snap judgment wrong. I appreciated then for the first time the difference between the world of business, where a 20 percent decrease in cost is a major triumph, and the world of science, where nothing seems worth doing unless it promises an improvement by a factor of at least ten. I hadn't done the calculations, because they obviously didn't permit such large savings. Alfred, on the other hand, considered it worth a day or two of his time to see if he could cut the cost of the magnet windings by $50,000.

By then, Loomis was convinced that Lawrence's project was of compelling importance to science, and he made it his mission to raise the money so that he could build the largest cyclotron ever made. Much in the same way Wood had once opened doors for him, Loomis arranged for the young Berkeley physicist to meet with many of the influential players in science, business, and philanthropy on the East Coast, personally squiring the wide-eyed country boy to a series of private meetings in Cambridge, New York, and Washington. Their rounds included Compton at MIT; Vannevar Bush, who had resigned as vice president of the university to become head of the Carnegie Institution of Washington, a private research organization founded by the steel baron; F. W. Walcott, the former senator and fellow Carnegie trustee, whose son had leukemia; and Frank Jewett, the head of Bell Labs. Compton in particular did a great deal to help carry the ball, bringing Lawrence and his proposal to the attention of the Rockefeller Foundation. After a morning session with Compton at his MIT office, Lawrence wrote Loomis to thank him for his help and update him on the prospective donors he was scurrying to line up:

> I had expected that both you and Compton would warmly approve the cyclotron project and would do everything possible to assist in bringing it about, but I did not expect there would be such widespread cordial approval, and it was a very pleasant experience to find wholehearted support in every quarter.
> Dr. Jewett telephoned a friend in Denver regarding Mr. Spencer Penrose; and, although we learned that Mr. Penrose was very ill with a cancer of the esophagus and could not be approached, it was arranged for me to go to Denver and consult his physician, who is also a close personal friend and adviser. This I did. Dr. McCressin received me very cordially, and after listening to our plans, said that he was sure that it was the kind of project that would interest Mr. Penrose very much and, should Mr. Penrose improve sufficiently in the near

future, he would be more than glad to put the matter to him favorably. . . . On the other hand, he said that in the event that Mr. Penrose does not get better his estate will go into the Penrose Foundation, which has in its charter the stipulation that the funds must be expended in the state of Colorado, thus ruling out any possibility of participating in this project.

The outlook for support in other quarters, however, is very promising, and, at the suggestion of Dr. Warren Weaver of the Rockefeller Foundation, we are going ahead immediately with the preparation of detailed plans. I am hoping that somewhere and somehow funds will be forthcoming early in the year so that we will be able to begin actual construction next spring. . . .

On November 9, it was announced that Lawrence had won the Nobel Prize for physics for his invention and development of the cyclotron. It came as something of a surprise, as for months there had been rumors out of Stockholm that there would be no Nobels awarded in chemistry and physics because of the war. Loomis was ecstatic. He immediately appreciated the weight that the award would carry in convincing those who still doubted the feasibility of Lawrence's monumental undertaking. The accompanying prestige and fanfare promised to make the job of fund-raising that much easier. Lawrence had ruled out making the trip to Sweden to attend the prize ceremony because of the danger of German submarines, particularly after his brother's close call, so when Loomis saw him a few days later, they celebrated his success with a drink.

As both Loomis and Lawrence were always impatient for results, it was their habit to conduct almost all of their business by telephone; therefore the correspondence between them tends to be hurried and spotty. But the latter part of 1939 and beginning of 1940 was a crucial juncture in the cyclotron project, and they had much to plot and plan, sending a flurry of letters back and forth between Berkeley and Tuxedo Park, where Loomis had returned in December for the holidays. Four days before Christmas, Loomis dashed off a triumphant note, addressing it to "my dear Dr. Lawrence":

Dr. Weaver spent a whole day with me, and we had a most interesting talk. He said he was looking forward to seeing you early in January. I was very encouraged with the conversation. I also spent quite a lot of time with Dr. Bush and others on the matter of the new machine, and I feel very encouraged. Karl Compton is coming down to spend a week with me in South Carolina in the middle of January, and I think his opinion will be of the utmost importance. . . .

Lawrence sent a quick reply to say he was "mighty glad" to get the good news and that plans for the cyclotron were going splendidly. Weaver had urged him to be bold and to beware of presenting the Rockefeller Foundation with a project "on too small a scale." Lawrence had taken him at his word and enlarged his vision yet again. "We have worked out enough of the details of machine weighing in the neighborhood of 5,000 tons to permit rather complete drawings to be made," he wrote Loomis enthusiastically. He promised to send along a copy of the artist's sketch of the proposed cyclotron in a day or two, adding, "It certainly is a thrilling thing to behold. We are perfectly sure this outfit will produce 200 million volt alpha particles, and there is a reasonable chance that we will be able to push on up to 300 million volts. I can hardly wait for the day when these plans will be realized."

Lawrence, who was by now the recipient of a generous stipend from Loomis, was also writing to request additional funds for two former graduate students he wanted on his team at Berkeley. In a calculated appeal to Loomis' competitive spirit, he noted that he had just had a letter from MIT's Robley Evans saying that their cyclotron was ready to go except for some trouble tuning up the oscillators, and that his former graduate student, Stanley Livingston, had reported that Harvard's cyclotron was running very well. Lawrence prompted, "To save further delays, it might be advantageous for Livingston simply to duplicate the Harvard radio-frequency layout. Incidentally, Evans wrote me that he hoped to come out here to work a while in the laboratory next summer if he can raise the money for his travelling expenses, and it seems to me a fine idea if travelling expenses would be provided for both Livingston and Evans."

He also inquired whether Loomis would "care to step in and help out" in another situation: Dr. S. Mrozowski, a member of the faculty of the University of Warsaw, had spent a sabbatical year at Berkeley before returning to Poland to take charge of the cyclotron project there and had fled Warsaw only a few days before the war broke out. Lawrence explained that he had "landed in this country with only a few hundred dollars" and had written to him of his plight. He suggested that $1,000 would make all the difference. "[He] is a good man and is proving a help in the laboratory, particularly now as we are short-handed; so it is entirely justified, quite apart from humanitarian aspects, to pay him to be with us . . . if it is entirely agreeable with you to provide a stipend (of possibly $150 a month for six months) for Mrozowski, it would certainly be appreciated." He concluded by sending his regards to Compton and company. "I wish it were possible for me to drop in on you some evening and sit and talk with you all, probably around a cheery fire in your Carolina island."

On January 5, Loomis sent a hastily scribbled note on Honey Horn plantation stationery explaining that he had asked his treasurer to issue a check for $1,000 from the Loomis Scientific Institute to take care of Mrozowski and assured Lawrence that he would speak to Compton so that Evans could "go out next summer."

When Weaver arrived in Berkeley on January 7, 1940, he was flabbergasted to discover that Lawrence's plans now called for a magnet weighing between four thousand and five thousand tons, in a housing 120 feet in diameter, and instead of the previous estimate of $750,000, he now wanted as much as $2 million. Lawrence was either blind to Weaver's dismay or overly confident of his persuasive powers. He may also have chosen to gloss over any nagging doubts he may have had, though it was not Lawrence's nature to dwell on the negative. In a long, ebullient letter to Loomis on January 13, Lawrence reported that Weaver's visit was "a great success all around. From the moment of his arrival it was clear that he was very keen for the project; and, as his visit here progressed, he became more enthusiastic. . . ." Weaver had also disclosed the confidential information that Compton had been nominated as a trustee of the Rockefeller Foundation, to go into effect in April. "It goes without saying that he will be a tower of strength on the board for the project."

Lawrence had managed to beguile Weaver with his eloquent justifications for why more than $2 million should be spent on his new cyclotron, but the others on the board of the Rockefeller Foundation were stunned by his extravagant demands. Lawrence's project had been growing progressively over the months, and as the physicist Robert Cornog observed at the time, if built to scale, it would certainly qualify as "the eighth, ninth, tenth and eleventh wonder of the world." The trustees concluded that Lawrence had let the California sunshine go to his head. He was asking too much. It was too risky. They lost their nerve, and as Weaver admitted during a conference with Loomis, the outlook looked grim that they would part with as much as half a million dollars.

Over the next two months, Loomis worked furiously to restore their flagging confidence. Compton did his part, sending a glowing letter of recommendation: "I not only consider Professor Lawrence's cyclotron project as 'one of the most interesting, the most potentially important, and the most promising projects in the whole present field of natural science,' but I should definitely place it in the number one position by a large margin among the various scientific projects of which I have knowledge at the present time. . . . No one could possibly question the selection of the University of California and Ernest Lawrence as the institution and the scientist to whom the project should be entrusted." On a more personal note, he added:

It happens I spent all of last week with Mr. Alfred Loomis on his South Carolina island discussing various problems involved in the construction of this proposed big cyclotron. We were interested first in trying to understand as thoroughly as possible the nature of the problems which would be involved in building a big machine, and second, in assuring ourselves that there were no inherent impossibilities in going to a cyclotron of so large a size. The net result of this study was our conclusion that there seems to be no insurmountable difficulties. . . . It is evident from this letter that I am not writing as a "disinterested party." However, my interest is entirely impersonal and has nothing to do with my own institution, and exists only because of my enthusiasm for the whole cyclotron project and my faith in Ernest Lawrence himself.

In early 1940, Loomis returned to California. A few weeks later, on February 29, the Nobel citation and medal were awarded to Lawrence at a ceremony on the Berkeley campus, with the Swedish consul making the presentation in place of the king. The university's president heaped praise on Lawrence and proudly declared that he had "discovered a blasting technique far more potent than anything Alfred Nobel ever dreamed of." The large party that followed at Di Biasi's restaurant in San Francisco was packed with all "the boys" from the Rad Lab and featured a cake shaped like the sixty-inch cyclotron, with "8 Billion Volts or Bust" written in colored icing. Lawrence received congratulatory telegrams from all over the world. Lee DuBridge, head of the Physics

Department of the University of Rochester, sent a limerick that was posted on the blackboard of the lab:

> *A handsome young man with blue eyes*
> *Built an atom-machine of great size,*
> *When asked why he did it,*
> *He blushed and admitted,*
> *"I was wise to the size of the prize."*

One of the most prescient notes came from Loomis' friend and mentor. Wood, by way of an old pioneer recognizing a new one, wrote Lawrence: "As you are laying the foundations for the cataclysmic explosion of uranium (if anyone accomplishes the chain reaction) I'm sure old Nobel would approve."

Apart from wanting to be in the thick of the action, Loomis remained in Berkeley primarily to pump up support for the cyclotron project in the last few weeks before the Rockefeller board convened. On Thursday, March 28, he met with Vannevar Bush. The following day, Loomis had organized a meeting on the second floor of the Rad Lab to discuss the project in detail with Lawrence, followed by an informal get-together that weekend at the Del Monte Lodge at Pebble Beach. He had invited a number of influential intermediaries who would be in a position to swing the Rockefeller vote in their favor. Karl Compton was coming out, as was his younger brother, Arthur Compton, a respected physicist, along with Harvard president Jim Conant. Cooksey captured the group in a photo in Lawrence's lab. "Alfred set up a show on Pebble Beach, where he had hired a whole darned hotel and gathered all sorts of people who had control of the money," recalled Bush. "He put on the show to back up Ernest's next venture, quite successfully, it goes without saying."

"They were all out there more or less as my guests," recalled Loomis. "I had a big party down in Del Monte. We did a great deal of work for radar, where we could get together uninterrupted." Loomis took advantage of his captive audience for a last pitch "about the necessity of the money for this big machine." He was also gambling that his prize winner's charisma would work its usual spell on all those assembled: "You can't get a group together for a long weekend without Ernest's affect on them." By Sunday, "there was no opposition."

On the following Wednesday, April 3, Weaver phoned Lawrence

that the trustees had awarded him $1.15 million, an astounding sum under the circumstances. This did not include the matching pledge from the university of $85,000 a year for ten years. Or the additional $50,000 Loomis helped angle from the Markle Foundation, which was to be one of several large donations he would secure for Lawrence to the tune of several hundred thousand dollars. According to Alvarez, it was the Tuxedo millionaire who made certain the pot of gold was at the end of the Rockefeller rainbow: "Loomis had been instrumental in securing the virtually unanimous support of the 'scientific establishment' for the proposal, thus relieving the Rockefeller Foundation of any necessity for acting as a judge between factions competing for the largest funds ever given to any physics project."

No sooner had the Rockefeller money been banked than Loomis whisked Lawrence off to Wall Street, where he spent two weeks making sure his boy got the best deals possible in purchasing the large quantities of iron and copper he required to build his great cyclotron. Loomis introduced Lawrence and Cooksey to his friend Philip Reed, the head of General Electric, who agreed to make the 184-inch cyclotron's power supply at cost. He then tried to induce Westinghouse to make them a better offer, to no avail. Loomis used his influence with Luis S. Cates, president of Phelps Dodge, to extract a promise that the increasingly scarce copper would be available, and he leaned on E. T. Stannard of the Kennecott Copper Company. He also took them to see Sloan Colt, head of Bankers Trust. "I knew most of these people very intimately, and it wasn't done on any official basis," he said later. "It was done on a more man-to-man basis," which was the way he and Lawrence did most things.

In Tuxedo Park, Loomis hosted a grand dinner for Lawrence and invited many prominent friends in the financial community who might be in a position to help him. The party was a great success, but while the after-dinner drinks were being served, one banker turned to Loomis and inquired under his breath, "Didn't the scientist come?" Telling the story later, Loomis explained that the charismatic physicist was so much like everyone else, so like the other successful businessmen in the room, that no one realized the Nobel laureate was at the table. Lawrence was hardly the owlish, white-haired professor they had expected. Always a great hit with women, Lawrence paid a great deal of flattering attention to the wives that night, particularly the pretty ones. "He loved to dance with them and in a very masculine way flirt with them," recalled Loomis. "He even at times made mild passes at them."

All in all, Loomis managed to get all the materials and equipment Lawrence needed for bargain rates despite the wartime shortages. He made sure every one of his friends all made good on their words and delivered the orders as promised. That it would turn out afterward that he was either a shareholder or had a controlling interest in several of these companies came as no surprise to those who knew him well. Loomis, of course, never mentioned it. On his return home, Lawrence regaled Alvarez with stories about his adventures on Wall Street with Loomis. Not one to be dazzled, Lawrence was nonetheless astounded at the ease with which the millionaire commandeered the city's resources:

> After spending some time with the Guggenheims, during which a favorable price for copper was negotiated, Loomis said, "Well, now we have to go after the iron. I think Ed Stettinius is the right man." [Stettinius was then chairman of U.S. Steel and later secretary of state.] Lawrence was impressed when a call was put through and Loomis said, "Hello, Ed, this is Alfred, I have somebody with me I think you'd like to meet. When can we come over?"

On another occasion, they went to see Seward Prosser, who was chairman of the board of the Bankers Trust Company and a trustee of the Markle Foundation. They were sitting in his office talking about Lawrence's research when Prosser turned abruptly to Loomis and asked, "Alfred, won't you come back and be on our board of directors? We greatly miss your counsel and advice."

Loomis smiled and, shaking his head, replied, "Seward, thank you very much, but I am certainly not coming back to your board of directors. I am now doing a much more interesting and important job."

As they left the Bankers Trust Building, Loomis told Lawrence that seeing his former colleagues again in this way, at that moment, was an extraordinary experience: "Here I am back in Wall Street getting the help of my old friends on a project that is dramatically constructive and forward looking, while [they] are here trying to hold together a tottering [financial] structure."

Loomis' masterminding of the Rockefeller grant for Lawrence's cyclotron did not go unnoticed. The way he overcame obstacles and brokered alliances, never taking no for an answer, made it clear just how effective a behind-the-scenes force he had become. Writing to Lawrence to congratulate him on the Rockefeller vote, Conant put in a plug for Harvard's cyclotron and gingerly broached the subject of recruiting the multitalented Loomis, who after all was an alumnus and seemed to have both energy and resources to spare: "I am still exploring the possibility of stealing back from you, if I can persuade him to come to us, our mutual friend who is a volunteer now in your outfit. It seems to me there is a possibility of using him, at least as a consultant, on both shores on the continent."

Lawrence sent Conant an obliging response: "I know that Alfred Loomis is very anxious to help you in any way he can. I need not say to you that not only is he a man of great ability, but also of fine character who is anxious to be useful in any way he can. Needless to say, we are glad to share our blessings, and fortunately for us it is possible for him to be very active in behalf of the work here and at the same time be very helpful in Cambridge." Loomis would indeed go to bat for the crimson team, lobbying the head of the Markle Foundation about making a grant to Harvard's cyclotron project and contributing his own "generous gift," which, as Conant wrote in gratitude, "will just make the difference between seeing the cyclotron work through next year." Conant sent off a note to Lawrence thanking him for "arousing Mr. Loomis' interest," adding, "I am sure you have helped us more than you realize."

That spring, at a luncheon honoring Lawrence at the Bohemian Club in San Francisco, attended by the scientific elite on the West Coast, Winthrop Aldrich, a member of the Rockefeller Foundation board, made a speech praising all that Lawrence had accomplished. In doing so, he said: "You may believe that the war in Europe is the most important thing that is going on for the human race at the present time, but actually it may prove that the events which are about to take place on the University of California campus may be of much more far-reaching significance to humanity." To the physicists in the audience, Cooksey wrote Loomis afterward, "it sounded like a reasonably fair introduction."

While he would always try to minimize his contribution, years later Loomis acknowledged that he was probably more conscious of the need to make haste and may have played an important role in persuading everyone involved that the cyclotron could not be built on the usual painstaking scientific timetable—that "science requires it, and requires it as fast as possible." Loomis, far more than Lawrence, was acutely aware of the imminence of America's involvement in the war. He had been working simultaneously on the cyclotron and the radar projects and had traveled to London several times to educate himself on Britain's radar technology, which was considerably more advanced. He had learned from his British colleagues just how depleted their industry resources were and how desperately they needed the United States' support.

"It was the early part of 1940," Loomis recalled. "Stimson was actually saying the war was coming very soon, and I'd been so close to him, and he felt so strongly about it." He had worked so quickly to get the money and supplies for the cyclotron because he knew that with the country spending so much money to rearm itself, and the situation in Europe deteriorating, a "business and raw material bottleneck" would soon cause delays. "I was a great help in speeding up the situation," Loomis would concede years later. "And those were the months in which weeks counted."

Chapter 8

ECHOES OF WAR

> Locusts hummed metallically in the distance, and the horizon
> wavered under a cloudless sky. Externally, the Laboratory was
> an idyll of summer peace.
>
> —WR, from *Brain Waves and Death*

B Y early 1940, Loomis was so caught up in his work with Lawrence
that he had decided to purchase a home in California and accept a position as a research scientist at the Rad Lab. The experience of working at
the Berkeley lab was intoxicating for Loomis. The sixty-inch cyclotron,
after numerous delays, was finally up and running, and it was a marvelous sight. With its big, powerful magnet, it promised to be productive
of exciting discoveries in nuclear physics, and Loomis could not resist
the chance to be a part of it all.

He had also decided that his next big project would be to back the
new research being done by Enrico Fermi, the brilliant young professor
of physics at Columbia who had won the Nobel in 1938—a year before
Lawrence—for his discovery of new radioactive elements and his related discovery of nuclear reactions brought about by slow neutrons.
Fermi happened to be in Berkeley from January 30 to February 20 as a
visiting lecturer, and Loomis, who had known the Italian refugee physi-

cist for several years, was quite taken with him. Fermi was enormously talented, with a lively and highly systematic mind, and he was open and generous like Lawrence. By his own account, he had taken up "the uranium split business with which half the world seems to be occupied . . . as soon as the cyclotron gave a beam."

Loomis was determined to help advance Fermi's work and arrange funding for him to build a nuclear chain reactor. Fermi's proposal was still on the drawing boards, but Loomis, who was always quick to seize on the next new thing, offered to help underwrite the cost of any experiments that might determine how fission could be exploited for atomic energy. If the explosive power of fission could be realized, it would give mankind command of an almost limitless supply of energy. With characteristic confidence and enthusiasm, Loomis plunged ahead into nuclear physics. A scientific race had begun, one with incalculably high stakes, and he was prepared to devote all his efforts and resources to developing the new field as fast as possible.

For Loomis, the discovery of fission by German physicists in the beginning of 1939, in light of the events in Europe, was both exhilarating and profoundly disturbing. The discovery was of great importance to science, yet such an enormous release of energy, if harnessed and controlled, would be a terrible weapon in the hands of Hitler. He had first heard the momentous news from Niels Bohr, upon the Danish physicist's arrival in the United States on January 16, 1939. Fermi had been at the pier to meet the SS *Drottningholm* to welcome Bohr, whom he and Loomis had recently visited in Copenhagen, and had promptly driven him to Tuxedo Park, where they both stayed for several days before going on to Washington for the Fifth Conference on Theoretical Physics. Bohr had barely set foot on American soil before announcing that the fission of uranium had been demonstrated by Otto Hahn and Fritz Strassmann in Berlin.

While he was in Tuxedo Park, Bohr had received a cable from the radiochemist Lise Meitner, a close associate of Hahn's, that she and her nephew Otto Frisch had confirmed the fission process in their own experiments in Sweden and that they had good reason to believe that when the uranium isotope 235 was bombarded with neutrons, it split into two lighter elements with a loss in mass and an enormous release of energy. "He got a cablegram from Meitner, and they thought it was 235 that was doing the splitting and that energy was coming out of it," recalled Loomis. "And then a week later he delivered the lecture before the National Academy, and practically before the sun was set it was confirmed in three labs in America."

The news that the experiments had been verified in American laboratories generated so much excitement that Bohr and Loomis had been able to talk of little else. Hans Christian Sonne, a Danish banker living in Tuxedo Park who had gone to boarding school with Bohr's brother and was a friend of the family, recalled that one evening after drinks at his house, Bohr and Loomis sat around discussing what the new development might mean for physics. "My father was a businessman, so he did not understand the scientific details," said his son, Christian Sonne, a real estate executive in Tuxedo Park, "except that it might make it possible to produce enough power to make a mighty big bomb."

Bohr's news had spread quickly, and when Leo Szilard heard about uranium fission a few days later from his friend Eugene Wigner, a physicist at Princeton, he was stunned. It was what he had been working toward for years. "When I heard," he recalled, "I saw immediately that these fragments, being heavier than corresponds to their charge, must emit neutrons, and if enough neutrons are emitted in this fission process, then it should be, of course, possible to sustain a chain reaction. All the things which H. G. Wells predicted appeared suddenly real to me." Szilard had immediately understood the implications of such a discovery. Europe was on the brink of another world war, and fission— which might be used to create "violent explosions"—had to be kept from the Germans. His first thought had been to contact Fermi and those physicists in Europe who were most likely to intuit this possibility and begin to organize self-imposed censorship on all nuclear research.

But once Bohr and Fermi spoke publicly about uranium fission in Washington, the cat was out of the bag. At the conference, Fermi, who had himself nearly discovered fission several years earlier, theorized that when a neutron knocked uranium apart, more neutrons might be emitted. He suggested there might be the possibility of a chain reaction— the release of atomic energy—and a bomb. In the days that followed, physicists everywhere rushed to their laboratories to test the process of uranium fission, and within forty-eight hours the key experiments had been replicated in several laboratories, including the Rad Lab, the Carnegie Institution, and Johns Hopkins. Bohr's information was published in the *Physical Review*, which later reported verification by Fermi at Columbia. Following a demonstration of uranium being bombarded with neurons at the Carnegie Institution on January 28, the *Washington Evening Star* carried a front-page story with the banner headline "Power of New Atomic Blast Greatest Achieved on Earth."

Throughout that spring, Bohr and Fermi continued to appear together at various scientific meetings, and their increasingly candid views on uranium fission, and its potential military use, were the talk of scientific and political circles. In mid-April, Bohr again stayed at Tuxedo Park for the weekend, where, as Loomis wrote Vannevar Bush, "he gave us a very interesting talk" on chain reactions with uranium and other heavy elements. After the American Physical Society's spring meeting on April 29, *The New York Times* wrote that the conferees argued "the probability of some scientist blowing up a sizeable portion of the earth with a tiny bit of uranium." Over the course of that summer, the intense activity and concern in the world of physics prompted the Einstein letter to Roosevelt warning him of the seriousness of atomic weapons. The whole matter came to a head in the fall of 1939 with the formation of a uranium committee, under the chairmanship of Lyman Briggs, director of the National Bureau of Standards. The Briggs Advisory Committee on Uranium was to be a panel made up of physicists and representatives of the army and navy and would coordinate secret research on a fission explosive. If a bomb was possible, and it unleashed enormous power, it would render the totalitarian war machines of Hitler and Mussolini unstoppable. It was essential that America's scientists and military organizations stay ahead on nuclear research should it ever prove feasible to build such a device. "Shortly after that," recalled Loomis, "the thing went underground."

As Europe slipped deeper into the war, the uranium panel twiddled its thumbs. It was so mired in bureaucracy that by the spring of 1940, it had managed to approve only the $6,000 in research funds earmarked for Fermi and Szilard, so they could purchase uranium and graphite for their fission experiments. A number of leading scientists were increasingly alarmed by the government's inaction, and chief among them was Vannevar Bush. During the weekend get-together in Del Monte, they had discussed the various possibilities for destruction inherent in fission. At the time, Bush had relayed the expressed concern of British researchers that a fission bomb could be developed. If the Nazis were to succeed first, they would control the world. There was general agreement that there ought to be a preparedness policy. Bush let it be known informally that he was working on a plan. He believed it was vital to find a way to organize the best brains and experts in the country to assist in the accelerated war effort and to help adapt the armed forces to the needs of a highly technical contest.

W H E N Loomis returned east in April 1940 to help Lawrence navigate Wall Street, he was more determined than ever to dedicate his private resources to scientific problems that might have value for defense purposes. Convinced that the United States would inevitably be drawn into the war, he was juggling several disparate projects related to mobilization and believed that priority should be given to things that could yield results in a matter of months or, at most, a year or two. Impatient with the MIT group's slow progress, he decided that the Loomis Laboratories would no longer muck about with a preliminary long-range exploration of propagation problems. Instead, it would focus on one pressing problem and work to find a practical and efficient solution.

While he was in San Carlos, Loomis had observed some early detection experiments done with a makeshift system that had actually been designed for blind landing tests. Even so, in the course of several tests with the ten-centimeter klystron and a ten-centimeter "Barrow horn" (a galvanized iron hollow cylinder used for transmitting ultrahigh frequencies, named after MIT's William Barrow), they had actually been able to pick up automobiles and trains a quarter of a mile away, using only a crystal detector with an audio amplifier. Loomis had also observed some of the experiments Hansen and the Varian brothers were performing with one of the first continuous wave Doppler radar sets. Impressed with what he had seen, Loomis proposed to develop a similar airplane locator based on the principle of the Doppler effect. He had brought back several ten-centimeter klystron tubes from California, and he had even convinced Hansen to come back with him to help assemble the setup in Tuxedo Park.

Loomis went up to MIT to arrange for several members of Bowles' team to be loaned to his laboratory for the summer. At the same time, he wrote Compton another check to help keep the university's ultrahigh-frequency program going. By the end of April, he had assembled a group consisting of Hansen and two young MIT graduate students, Donald Kerr and Frank Lewis, and together they made the drive from Cambridge to Tuxedo Park. There they were joined by MIT's William Tuller and William Ratliff, now with the Sperry Co. At first, Lewis did not know what to make of Loomis and his lavishly equipped laboratory, and in telling the story later, he joked about his utter astonishment at the exclusive surroundings he suddenly found himself in: "Now this Tuxedo Park is a private enclave where people go who don't want to be bothered with other people just driving in and saying 'hello.' They have a fence around it and they had a gatehouse where you go in and check yourself through. Everybody who was run in and out of there was thoroughly understood by the people that opened the gate. So if they didn't know you, you didn't get in."

As they settled down to work, Lewis began to do "a little inquiring" about the mysterious millionaire who was bankrolling their operation. An enormous amount of equipment was needed, and Loomis "footed the bill generously," as well as paying the salaries of himself and the other newcomers who were not covered by the original grant. They were joined by several of Loomis' longtime associates: Garret Hobart; Charles Butt, who was E. Newton Harvey's research assistant at Princeton and a regular during the summers; Philip Miller, the lab's manager, machinist, and jack-of-all-trades; and Loomis' youngest son, Henry, who was in his third year at Harvard but had enlisted in the navy and was due to ship out in six weeks. There was also an impressive stream of visitors, including R. W. Wood, who filled Lewis in on Loomis' background, how he had made his money, and his close ties to Henry Stimson. As the weeks went by, Lewis came to have a grudging respect for what his wealthy host was trying to accomplish on his own dime:

"Loomis was anxious to have people there who knew about microwaves because he had a feeling microwaves were going to be important," explained Lewis. "He was a person who loved to be with the leaders of any one particular enterprise. As such he was called a dilettante by people who thought that was a good name for him. So you had to work for him and talk to him a bit and then you found out there wasn't anything phony about him at all. He was a first-class scientific person, and he had a lot of money. With those two things he could do a lot of things."

Aggressive and enthusiastic, Loomis insisted on getting started right away. His time on the East Coast was short, and he had to be back in Berkeley before too long. In any case, he was a hands-on experimenter and believed they stood a much better chance of accumulating useful data if they had a system in operation. It was the way he and Wood had always worked, and it was the way he intended to proceed now. They would just have to make improvements on the fly and incorporate new equipment as it became available. Bush, who had kept a close eye on Loomis' project at Tuxedo Park, felt the work was promising enough that he had told Ed Bowles, the MIT radar expert who had been collaborating with Loomis on microwave research, if he "needed further support, to let me know." Bowles did, and Bush directed the Carnegie Institution to allot the sum of $10,000 for Loomis' microwave detection project at Tuxedo Park.

In the beginning of May, Loomis received a letter from Cooksey telling him that Lawrence was down with a "terrible infection" and updating him on the contracts for the 184-inch cyclotron. Cooksey included a bulletin about a possible breakthrough in radar technology: "Professor Marshall of the Electrical Engineering Department, who, you will remember, is working with Dave Sloan on the micro-wave tube, just back of the partition of my office, told me Friday of some of their successes. . . ." They had tested a tube that oscillated at 50 centimeters and at 2,500 watts and were now in the process of running more checks. It looked as though a piece of cyclotron technology had been transformed into a generator of radio waves at a frequency and power suitable to radar. Loomis was elated by the news. It meant that radar sets could be smaller, while the detail of what they could see increased. The Sloan-Marshall tube, or "resnatron," could well become the basis of a new generation of powerful, airborne microwave transmitters.

At the end of the letter, Cooksey added that the physicist Emilio Segrè, who was attached to the Rad Lab, had returned from New York and "tells us there is a great deal of hush hush around Columbia in regard to the work that is proceeding on uranium isotope separation and uranium bombs. Apparently the army and perhaps the navy are interested. I have no words with which to express my feelings about the conditions in Europe. I merely know that it is what we must expect. . . ."

Loomis responded with a long letter outlining the status of his various negotiations for the tons of steel laminates for Lawrence's cyclotron. He confirmed that "U-235" had indeed become "very hot" and that Bush was organizing a conference in Washington the following week for all the scientists who were working on it. As for his microwave work, he wrote, "This whole problem has speeded up enormously because of its immediate war demand" and promised more details later. In a subsequent letter to Lawrence, Loomis wrote that he was working on the Sperry people to provide financing for Sloan's research. "If the tube is as promising as it seems to be, there ought to be no delay in pushing it." The Sperry company was building a large factory in Hartford and had bought a private airfield and might be using his Tuxedo facilities as well. He had put Bowles "in charge" of the laboratory. He added that while he hoped to return to California with Lawrence after his planned visit to Tuxedo in late June, "I don't know how long I will be able to stay out there this summer, if this micro-wave work gets very intense." It was the first hint Loomis had given Lawrence that he might no longer be able to continue carrying on his share of the responsibility for getting the big cyclotron built. Putting his duty to his country first, Loomis had put his plans for a new life out west on hold.

During this same period, Loomis had a visit from his old friend George Kistiakowsky, who wanted to pass along some troubling information that had come his way. Kistiakowsky had heard that the Germans were carrying out extensive uranium research at the Kaiser Wilhelm Institute in Berlin and feared that with their advanced laboratories and aggressive approach, they might be the first to develop a fission bomb. Kistiakowsky was well aware of Loomis' reach into the highest levels of science and government and felt sure he would see to it that the facts were communicated to the right people. Loomis decided every attempt should be made to discover the status of the Germans' uranium experiments, and he enlisted Kistiakowsky to use his network of émigré scientists and European colleagues to gather information and offered to pay for his time and expenses. He then arranged a meeting with Compton and told him what he had learned from Kistiakowsky. On May 9, Compton wrote Bush a confidential letter apprising him of his conversation with Loomis:

As far back as the first suggestion that uranium fission might have a very significant industrial or particularly military significance, he [Loomis] has been very close to groups involved in his work. . . . It is now clear that the German scientists are concentrating major efforts on this problem at the Kaiser-Wilhelm Institute. It also appears clear that uranium 235 could be a tremendously powerful war weapon if it could be secured in substantial quantity and a fair degree of purity. Alfred makes the pertinent suggestion that we really ought to get together some of the most competent men in the field to analyze the possibilities in the situation and be ready to proceed actively if a promising program develops. . . . George Kistiakowsky is intensely interested in the subject (as a weapon which must not be allowed to develop first in Nazi hands) and he is also making an independent study and will report to Alfred Loomis. . . .

Bush promptly wrote to Loomis in Tuxedo Park on May 13:

The matter of uranium fission is exceedingly active. I rather think that the [Carnegie] Institution should take the lead in furthering and correlating this whole matter. Dr. Jewett and I are to see the Army and Navy today on the subject. If we proceed actively we will of course have to use some of our [MIT] Corporation grant, and this matter will probably come up, therefore, at the meeting of the Executive Committee which follows the Finance Committee meeting on the twenty-third. If it is going to be possible for you to be about, I would suggest to Governor Forbes that you be invited into the Executive Committee meeting when the matter is considered. I will be able to give you further information on this particular subject before very long. . . .

On May 17, Bush sent Loomis a quick note informing him that the governor had authorized him to attend the executive committee meeting, "inasmuch as you are so well informed on the subject."

Loomis, as always, kept Lawrence abreast of the latest news on the uranium front and the behind-the-scenes effort to organize a campaign to move public opinion in favor of taking advantage of modern scientific developments for military purposes. As Lawrence wrote Loomis on May 20, "I gather from your letter that something along this line may be imminent. We in this country certainly do not want to miss any bets in this direction. Incidentally, Mr. Rockefeller would not like to hear this, but we certainly will not be unmindful of the possibilities of discoveries of military value in the energy range above 100 million volts. I am betting that we will find all kinds of fission reactions in many of the heavy elements."

Three days later, Bush convinced the Carnegie Institution's executive committee to give him $20,000 "for a defense research project concerning uranium fission." Even though he shared the prevailing opinion that an atomic explosion was "remote from a practical standpoint," he agreed that they should push ahead on exploring the possibilities: "I wish that the physicist who fished uranium in the first place had waited a few years before he sprung this particular thing on an unstable world. However, we have the matter in our laps and we have to do the best we can."

As far as Bush was concerned, the Carnegie money was only a drop in the bucket. He believed it was vital that science and technology were broadly mobilized for the war, which would provide him with a way to address what he saw as by far the most pressing military problem—the need to rapidly improve the country's air defenses. He was convinced that airpower was the backbone of military strength. America, which had been isolationist since the First World War, was vulnerable only to transatlantic attack. Radar held the key to revolutionizing warfare by providing a better means to track the enemy and accurately destroy targets. But to date, the army had ignored radar's potential for defensive action and could not be interested in sponsoring any research. The navy had developed its own detection devices but was woefully short of funds to do further research. As Bush had complained to former president Hoover in a letter a year earlier: "The whole world situation would be much altered if there was an effective defense against bombing by aircraft. There are promising devices, not now being developed to my knowledge, which warrant intense effort. This would be true even if the promise of success were small, and I believe it is certainly not negligible. . . ."

The "intense effort" Bush had in mind would require unprecedented cooperation among three distrustful—at times even hostile—communities: military, science, and industry. It would also require an enormous amount of money. But Bush believed with the right leaders working together, he could create a new military research organization that could exploit the technical advances that were indispensable to modern warfare. "I was located in Washington, I knew government, and I knew the ropes," he would say later. "And I could see that the United States was asleep on the technical end." As head of the Carnegie Institution, he was held in high regard by the scientific establishment, and he knew he could count on the backing of four men who figured prominently in its ruling echelon: Compton, Conant, Jewett, and Loomis. Compton and Conant headed major universities; and Jewett was president of the National Academy of Sciences and Bell Labs, easily the most respected industry research center. Loomis was a well-connected banker, but in

Bush's view he had gained acceptance into that brotherhood by virtue of his "real contribution to scientific knowledge." What drew these men together, he explained, was "one thing we deeply shared—worry."

On June 12, as the German blitzkrieg attacked the French country-side, Bush went to see Roosevelt to make his case for the creation of the National Defense Research Committee (NDRC). They were joined in the Oval Office by Harry Hopkins, the president's closest aide. Bush presented Roosevelt with a single sheet of paper clearly spelling out his plan for mobilizing the scientific community for war. The document, containing a brief four-paragraph outline of his proposed agency, stipulated that the NDRC would be made up of members from "War, Navy, Commerce, National Academy of Sciences, plus several distinguished scientists and engineers, all to serve without remuneration." Its function would be "to correlate and support scientific research on mechanisms and devices of war." The NDRC would work in close liaison with the military, but independent of its control.

After months of feeling that his hands were tied by a campaign promise to protect American boys and adhere to the isolationist policy, Roosevelt jumped at the opportunity to take constructive action on another front. He had already had to decline countless requests from Britain's prime minister, Winston Churchill, and France's beleaguered premier, Paul Reynaud, calling for America to intervene in the war. If America would not come to their aid, it could at least supply them with arms. What was needed in the present emergency was a massive weapons production program, one that could be up and running in record time. Bush agreed to head up the program and was promised a direct line to the Oval Office and as much money as he needed from the president's special fund. It took only ten minutes for Roosevelt to approve Bush's audacious plan, scrawling, "O.K.—FDR," across the memo.

On June 14, German tanks rolled into an undefended Paris. One week later, France surrendered and was forced to suffer the humiliation of signing the agreement in the same railway coach in which the Germans had signed the armistice of November 1918. The photograph of a triumphant Hitler posing in front of the military convoy was on the front pages of all the newspapers. England braced for a bloody summer. "The Battle of France is over," Churchill told the House of Commons. "I expect the Battle of Britain is about to begin." Stimson, who was now seventy-two and had spent the past year enmeshed in a difficult and exhausting legal case, could no longer stay silent. On the night of June 17, he gave a radio address laying out seven steps that should be taken immediately in the national defense, beginning with the repeal of the "ill-starred so-called neutrality act." He also called for the opening of U.S. ports to British and French vessels for repair; the accelerated supply by every means in our power—if necessary, by navy convoy—of war matériel to England and France; and the immediate adoption of compulsory military service. The following day, he received a call from President Roosevelt asking him to become secretary of war.

Upon hearing that Stimson was entering the cabinet, Lawrence immediately wrote Loomis: "I know if I were Colonel Stimson I would certainly be depending on you for all sorts of advice," he predicted. "In fact, I would create a new job, Under-Secretary of War for Technical Matters, and draft you for the job."

I F anyone could mold a group of civilian scientists and skeptical military leaders into an all-out American defense organization, Loomis believed it was Bush. He was, by nature, an inventor, a fixer of problems and machines. "I'm no scientist, I'm an engineer," Bush would often say of himself, and the assertion reflected his impatient, hard-nosed Yankee demeanor. He had a long-standing interest in the applications of science to war that dated back to his days doing antisubmarine research in World War I. Throughout his career, first at MIT and then at Carnegie, he had straddled the worlds of basic and applied research and brokered deals between university and industry laboratories. Every bit as autocratic as Loomis himself, he was not afraid to break the rules or bend institutions to his will. There was a strong personal and professional bond between the two men and a mutual admiration for their ability to get things done. "Of the men whose death in the summer of 1940 would have been the greatest calamity for America," Loomis would observe a few years later, "the president is first, and Dr. Bush would be second or third."

Bush, in turn, held Loomis in extremely high regard. He knew of his reputation as a financial genius and knew firsthand of his keen scientific mind. According to Caryl Haskins, who worked for Bush in Washington during the war, the two men had an extremely warm and easy relationship, and when they conducted business it sounded "just like two friends having a conversation." For two such charismatic, larger-than-life men, they could both be very subtle. They thought along such similar lines at times that they almost finished each other's sentences, and along with a bone-dry sense of humor, they shared a love of tobacco. Loomis was always lighting up cigarette after cigarette from the pack of Lucky Strikes he always carried on him, and Bush was rarely without his pipe. "Alfred was the same-caliber man as Bush, and they recognized it in each other," recalled Haskins. "It was simply a great personal relationship. Alfred was more talented than most people, and he had a gift for talented people. Bush respected him, and knew his abilities well enough to ask him to come to Washington to help. And he did come."

Bush decided to divide the NDRC into five divisions, each to be headed by a member of the main committee. Each division would then be composed of several sections, which would serve as the true operating units of the organization. The section chairmen were therefore key and had to be chosen with great care. They would be respected civilian scientists and required to work at demanding full-time jobs on a voluntary basis. Bush wasted no time drafting his friends, appointing Conant as his deputy and tapping Jewett, Compton, and Loomis for top positions. He also called up Richard Tolman, the respected theorist of physical chemistry, who was head of the California Institute of Technology, and Conway Coe, the commissioner of patents. Together, they would serve as the top scientific generals in the coming war. It would fall to them not only to recruit scores of individual scientists, but also to write hundreds of research contracts with universities, industrial laboratories, and research centers. Each would have the power to vastly enrich his own institutions and friends, which under different circumstances would have given rise to all kinds of questions about conflicts of interest. But the crisis demanded that everyone act decisively and judiciously, and Bush trusted his team to rise to the challenge. No one was more vulnerable to charges of self-interest than Loomis, who had been a prominent Wall Street figure and enemy of the New Deal, and whose cousin, Henry Stimson, had just been named Roosevelt's secretary of war. Their "kinship," wrote Bush, "might easily have created a problem but for Alfred's care to avoid it." But the threatening world would make for strange bedfellows, and Loomis and Stimson were the first of many New Deal critics the Roosevelt administration would embrace in the months to come.

To deflect any criticism Loomis' appointment might attract from the ranks of professional scientists, Bush and Compton quickly mounted an effort to have his name put forward in the National Academy of Sciences, a private organization created to provide expert advice to the government, whose membership carried their cadre's ultimate seal of approval. Lawrence had suggested the idea to Compton in a letter only a few weeks earlier, no doubt at Loomis' behest:

I hope the microwave work at MIT is being accelerated, as doubtless it is. It is mighty fortunate that Van [Bush] is in Washington to influence the army and navy toward a more scientific approach to the problem of warfare, and I wish that Alfred now were a member of the National Academy in order that he could be more influential in that direction. . . .

With Bush and Compton in on the game, Lawrence enthusiastically championed Loomis' nomination, and as he promised Compton in a brief note on June 5, "I am going to make it a business to see each member of the physics section." He added that he would also like to lobby another section but was not sure what field was most appropriate, biology or engineering: "Alfred certainly is well-known to the biologists for all his work, but whether he is known to the engineers is not clear to me." He suggested Compton talk to various academy members and "find out how they feel about it." Bush, Compton, and Lawrence then turned to Jewett to help resolve the awkward matter of how best to frame Loomis' scientific contribution and make sure that it passed muster with the academy members. As the reluctant Jewett warned Compton, someone as eclectic as Loomis would not be an easy sell:

I am returning herewith, signed, the Intersectional Nomination for Loomis, which you sent me with your letter of June 20th. While there is no question Loomis would be a valuable addition to membership of the Academy, and while Bush and I have talked the matter over extensively, I had a bit of uncertainty as to whether it was best for me to sign because of my position as President of the Academy. However, by signing you will see I have resolved my own doubts.

As between the two proposals for nomination, I incline toward the two-section combination of Physics and Engineering because I believe it would be easier to qualify Loomis as an engineer than as a physicist. However, if he can be qualified as a physicist I am sure a large majority of the engineers will support him and it might make it easier getting in more border-field people. . . .

It took months of tireless "electioneering," as Compton put it, to get Loomis in under the heading of physicist. As he wrote Lawrence at one point, "My defense for presenting Loomis' case was simply that his activities were so much on the borderline of physics . . . that there was a danger of a man very valuable to the Academy being lost sight of because he fell betwixt and between formalized sections." Compton also warned Lawrence to step back from too much overt politicking: "I doubt whether it would be advisable for another letter to go out on behalf of Loomis but I do think that a little personal missionary work as the occasion offers, or perhaps a personal letter to a few members of the Section who are such close friends of yours that the letter would not be taken amiss, might be worthwhile. . . ."

In the meantime, Bush had arranged for the NDRC to be given jurisdiction over the Briggs uranium committee and various other subcommittees, all unpublicized. American scientists were asked to comply with wartime censorship and exercise extreme discretion, and scientific journals were instructed not to publish papers on fission and any related subjects. The NDRC immediately began to inventory the country's research facilities and technical manpower. Compton approached the military agencies to compile a list of critical projects, those programs not yet under way that would be worthwhile, and those needing to be supplemented. As the program began to take form, and Conant assumed more of the burden of administrating the massive effort, he was amazed by the autonomous and far-reaching powers granted to Bush and his deputies at the NDRC: "Scientists were to be mobilized for the defense effort in their own laboratories. A man who we of the committee thought could do a job was going to be asked to be the chief investigator; he would assemble a staff in his own laboratory if possible; he would make progress reports to our committee through a small organization of part-time advisers and full-time staff."

Bush himself later admitted he had pulled off something of "an end run, a grab by which a small company of scientists and engineers, outside established channels, got hold of the authority and money for the program of developing new weapons."

Over a period of weeks, responsibility was divvied up among the five main divisions, with Compton assigned to take charge of Division D—the radar division. Compton asked Loomis to be chairman of the special microwave committee (Section D-1). He netted the assignment in part because he had been immersed in the subject for the past year and quite possibly knew more about radio detection than anyone, and in part because of the close relationship he had forged with Compton and Bush. Moreover, Loomis had proven himself a gifted improviser in marshaling support for the big cyclotron, a man who could move mountains if they stood in his way. It also just so happened that he possessed one of the finest facilities in the world to carry out exactly this kind of work and was presently doing pioneering radar research at his Tuxedo laboratory in conjunction with MIT.

Loomis chose the members of his own panel, appointing Bowles as executive secretary. After contacting Ralph Bown of the Bell Telephone Company, and Hugh Willis of Sperry, he proceeded to hold his first meetings at Tuxedo Park on July 14, 1940, before any of them had received their formal appointments from Bush. They set out their administrative needs, methods, and objectives and decided for the sake of speed and efficiency that it was important to keep the group small. They defined their goal in clear terms: "So to organize and coordinate research, invention, and development as to obtain the most effective military application of microwaves in the minimum time."

From day one, Bowles chafed at working under a man he felt, for obvious reasons, was an interloper. Loomis' unorthodox and occasionally high-handed methods would further stoke his resentment. "The microwave committee was to me a kind of mongrel gathering at the start," Bowles recalled, placing the blame squarely on Bush's shoulders for approving it in the first place. "Loomis had been a rather devious financier, and an operator par excellence. You never knew what the hell he was up to behind the scenes. But he was a driver, and brilliant, too brilliant for his own good in my judgment." But the British had already established the noble tradition of using "indigent scientists" to help apply modern technology to the military need of the country. "In Bush's mind, as I read it," said Bowles, "there was in this country a vast body of highly trained minds in the field of science diffusely spread throughout the country in our educational matrix. Here was a resource scattered from here to breakfast, relatively inactive as it stood, which could contribute material were it organized and well directed."

Despite their differences, Loomis and Bowles managed to work together fairly well in the beginning. The committee was composed primarily of representatives from industry, including GE, Sperry, Westinghouse, RCA, and Bell Telephone. The army also had a presence, though it was there primarily to raise questions. It was an awkward mix, and not easy to steer, which made for lots of disagreements, according to Bowles, and "a lot of unhappiness." Early on, the committee members deadlocked on the issue of who should actually produce the radar sets. The corporate worthies felt the MIT scientists should be confined to basic research and kept finding reasons why their great industrial laboratories should be put in charge of the primary development and production. Loomis and Bowles both felt strongly that the bulk of the development work should be done in one dedicated laboratory, and they ought not to depend solely on outside contractors. To overcome the deadlock, a number of decisions were passed up to Compton and the NDRC, which sided with Loomis. When Jewett, the head of Bell Labs, found out, he felt he had been "double-crossed." This led to "a very hot meeting," recalled Compton, and some tension between various parties for some time thereafter. But had Loomis not held his ground, Bowles later grudgingly admitted, the radar project certainly would have ended up solidly "in industry's control."

The microwave committee's first priority was to determine what research the army and navy wished them to undertake and to award the contracts to the best candidates. During the summer months, Compton and Loomis toured all of the important radar developments, first paying a visit to the Naval Research Laboratory (NRL) at Anacostia, and then the army installation at Fort Monmouth, New Jersey, where Colonel (later General) Roger B. Colton, director of the Signal Corps Laboratories, showed them around. Because both the navy and army research projects were regarded as strict military secrets, no "outsiders" knew of them, so Compton and Loomis had to be officially informed of the significant technical advance known as pulse radar.

The basic principle of pulse ranging, which characterizes modern radar, had been around for some time and had been discovered almost simultaneously in America, England, France, Germany, and Japan. In the United States, it was first used in 1925 by Merle Tuve and Gregory Breit at the Carnegie Institution in Washington for measuring the distance to the earth's ionosphere, which is the radio-reflecting layer near the top of the atmosphere. The technique consisted of sending skyward a train of very short impulses, a small fraction of a second in length, and measuring the time it took the reflected pulse to return to earth. In 1933, it had occurred to scientists at the Naval Research Laboratory that the pulse technique could be used to detect objects such as airplanes and ships. Over the next few years, they solved the problems of generating pulses of the proper length and shape, developing a common radar antenna for both transmitting and receiving called a "duplexer," and designing cathode ray tube displays for the received pulses. By 1936, the army, working independently at their Signal Corps Laboratories, had invented a detector for use by antiaircraft batteries. The system not only detected radio pulses from aircraft, but passed on information about their direction, elevation, and range. Even though the British had lagged behind, they quickly took the lead in radio detection under the Scottish physicist Robert Watson-Watt, who first patented radar in 1935 for his meteorological studies and then, in an environment of impending war, quickly applied it to military defense.

By the time Compton and Loomis were being introduced to pulse radar, the navy had named their system "radar," a manufactured term that was an abbreviation of "radio detection and ranging," while the army referred to their outfit as RPF, "radio position finding." The British, meanwhile, called their closely guarded system RDF. As the war effort got under way, the more convenient term *radar* would be adopted by common consent by the U.S. forces and subsequently, in 1943, by the British. All the radar development projects were "shrouded with a terrific amount of secrecy," according to Compton, and they came away with the distinct impression that neither branch was aware of the research being done by the other service. Consequently, they "felt duty bound to avoid being a channel by which information could be conveyed from one group to the other." The third meeting of the microwave committee was held on July 30 in Washington. The night before, there was a big dinner at the Wardman Park Hotel attended by Bush, Compton, and Jewett, followed by an evening discussion in the large suite Loomis kept there for such purposes. Loomis introduced the senior army and navy officers to the various committee members, and another trip to review the radar equipment was arranged.

In late August, Compton and Loomis were asked to attend the army maneuvers taking place at Ogdensburg in upstate New York and were flown up in Secretary Stimson's private plane. They were there to observe one of the first field tests of the army's top-secret pulse radar system, the SCR-268. As Compton later recalled, they "were quite the envy of the high officers attending the maneuvers because not even the generals were allowed to get near enough the equipment to find out anything about its operation." During two days at Ogdensburg, they saw the chief test, which consisted of a comparison of an airplane detection by pulse radar and by a network of volunteer observers reporting in by telephone. A huge plotting board was set up, and the army had acquired priority phone lines for volunteer observers scattered all over the upper part of the state. On the big board, all planes reported by the observers were marked down and tracked as to the type of plane, location, and heading. A parallel plotting system was set up using information acquired by the SCR-268, which, of course, won the day. They were greatly encouraged by the results with the army radar sets, which detected a flight of army planes at a range of about seventy miles.

Loomis was largely responsible for the committee's wholehearted sponsoring of microwave radar research. The army was skeptical, believing that microwave radar "was for the next war, not this one." The army had already worked to improve its transmitting tubes so that the wavelength could be reduced to 1½ meters and thought anything much shorter than that could not be perfected anytime soon. Given how slow they had been to capitalize on new technology in the past, Loomis regarded the army's attitude as more of a reflection of their own bureaucracy than those posed by the research challenge. The opinion was also based on the peacetime experience of longer-wave radar and the fact that big companies like GE and RCA had been on the verge of shutting down their microwave work because they could not find any commercial applications. The navy viewed itself as the aristocracy of the armed forces and jealousy guarded its radar technology, making it clear that they wanted as little to do with Loomis' committee as possible.

But Loomis, from the outset, believed pursuing the field of microwaves was a matter of the utmost urgency. As usual, he was not afraid of advancing an idea that might be unpopular; and he was used to relying on his own counsel. Just before the Battle of Britain in July 1940, he had made a tour of the English radar research laboratories and learned that their most pressing need was for a microwave system for night fighters and antiaircraft guns. Loomis had even talked to the British about ways in which they might cooperate in the area of microwave radar, so that America could carry on the work started in the United Kingdom. Loomis knew from his own research that the chief obstacle to microwave radar was the lack of a vacuum tube capable of generating sufficient energetic radiation at such short wavelengths. Only two tubes held out any immediate hope of providing real power on wavelengths below one meter: the klystron, which they were working on at Tuxedo Park, and the Sloan-Marshall resnatron at Berkeley. Loomis was convinced that if they could just find a solution to this problem, it would lead to an enormous widening of the powers of radar.

Almost as soon as he took charge of the microwave project, Loomis called on Lawrence to help him advance the development of the Sloan-Marshall tube with "the utmost vigor." They desperately needed an oscillator to produce still shorter wavelengths and a satisfactory power source to go with it. "Can't you step in and take responsibility for organizing it in a large way and have it the major war research of the University?" he asked in a hasty letter on July 9. "If a tube of 25 to 50 kilowatts at 20 to 35 centimeters were available there are some very pressing problems that could be powerfully attacked." Loomis promised Lawrence $20,000 from the NDRC; raised $4,500 from the Research Corporation, which had patented the tube; and threw in an additional $1,500 of his own money. He concluded on a wistful note, worrying that he must be missing out on everything at the Rad Lab: "I have been thinking a great deal about the cyclotron, and I can't tell you how anxious I am to catch up on all the new developments. . . . [Ellen] was talking last night about how wonderful it had been out there last winter 'before the war.' "

Lawrence immediately wrote back that he would take the necessary steps to see that the project proceeded "full speed ahead" and, if need be, would draw on funds from his precious 184-inch cyclotron to get the job done. It was the first microwave contract Loomis would approve, and one of the very earliest of the war effort. Loomis would go on to ask for Lawrence's help on any number of other war-related devices, and their enthusiasm for invention and shared pleasure in pooling their imaginative ideas and practical skills are evident as they worked out their ideas for various ingenious gadgets. Taking a page right out of one of Wood's infamous investigations on behalf of the police, as in the bombing of Morgan Bank, Loomis soon invited Lawrence's collaboration on an important and mysterious "FBI problem" he had been approached about:

> Suppose the FBI or the Naval Intelligence Unit would like to mark certain confidential documents for the purpose of catching a person that they suspected of being a spy, especially in the case where such a person was so high up in the organization that they could not afford to make any false accusations and must have absolute and immediate proof. I suggested that a few drops of a radio active preparation could

be placed at the exit of the building or other suitable place and that if such a suspected person passed near the counter they would have conclusive evidence on which to arrest and search him, whether he carried a document on his person or in his briefcase. Could you let me know what substance you would suggest and whether you could supply some small amount for test. It should of course be a substance whose radiations could easily be distinguished from those coming from the luminous dial of a wrist watch. I think we can assume that the suspect would not know the method and would not provide a lead container.

Lawrence considered Loomis' solution to the FBI problem "thoroughly practical" and passed along his own inspired idea for a spy-catching device:

There are numerous radioactive substances suitable for such purposes. One that comes to mind immediately is radio-yttrium, having a half-life of 105 days and emitting a very penetrating gamma-ray . . . the radioactivity can be concentrated into hardly more than a pinpoint of material which would have its advantages if it were to be used for labelling a document. . . . I have gone ahead and assigned an assistant to Luis Alvarez to work out the design and build an extremely portable Geiger counter which could be worn inconspicuously on the person in order that a special agent might carry it and, by walking near an individual, determine whether he was carrying any radioactive material. I suggested that instead of an earphone to detect the Geiger counts that they develop an arrangement whereby the counts produce tiny electric shocks on the skin of the individual carrying the counter so that the arrangements could be kept completely out of sight. I think such a gadget might prove to be very useful.

Loomis heartily approved the idea and wrote back that he would like to pick up the portable Geiger counter and a sample of radio yttrium when he was in California the following month so he could demonstrate the idea before naval intelligence and the FBI in Washington. He added, "I do believe that after the battle for England starts, and after we have universal conscription, this country will appreciate more and more the gravity of our situation." After Loomis' trip west was postponed indefinitely because of the intensifying demands of the radar work, Lawrence's little vest-pocket Geiger counter, which he boasted was now so compact that it "takes about as much space as a New York gangster would allow for his guns," was shipped to Tuxedo Park. Loomis showed it to "several important people" in Washington and reported that it worked perfectly. But it seemed that "carrying it on the person is not so convenient," and he had given some thought as to how best to conceal it. He went on to elaborate his idea:

> It occurred to me yesterday that a book would probably be the most suitable. If a book was used the high tension source could probably be obtained in a smaller space by charging a group of condensers in parallel. This would involve pressing a button on the side from time to time, but that would not be serious, especially as in that method there would be no danger of a current drain on the batteries. Could you send me two or three of the counting tubes themselves? Would you also think over the best solution to put the radio yttrium in? I should think there would have to be some chemical compound that would unite with the paper in such a way as not to make too noticeable a stain, and yet would hold the yttrium in the paper when it was dry.

In the midst of all the urgent business at hand, the two physicists spent months writing letters back and forth detailing further refinements of the little Geiger counter, each adding his own whimsical embellishments. Lawrence pooh-poohed Loomis' book idea as impractical and instead reported that he would be sending a new outfit based on a design cooked up by Alvarez and his assistant. The new design "gets away from the mechanical vibrator and transformer and instead uses a tiny radio frequency oscillator from which the voltage was stepped up to about 900 volts from a 45-volt battery," Lawrence wrote, adding that it would be "compact enough to carry in one's coat pocket with the necessary batteries in one's hip pockets. Needless to say, the whole unit including batteries could very readily be incorporated in a book as you suggest." He also addressed Loomis' suggestion that they change their approach on another front—namely, the Sloan tube. Loomis had written about Sloan's idea of producing high-speed electrons by sending powerful waves down a hollow pipe that contained bulges at proper intervals. In this case, the energy of the waves was progressively transferred to the electrons. Loomis had requested they "think carefully of the reverse process, namely, sending high speed electrons down such a tube and taking out power waves at the other end. These waves might never have to see a transmission line and might go right on out from the generator to a hollow pipe to a horn." Lawrence exercised great care in replying to his brilliant but meddlesome friend, who was in the habit of telling everyone what to do:

> As regards your thought about reversing Sloan's electron accelerator and using it as a micro wave generator, it is, as I can judge, a good idea and completely feasible. I discussed the matter with Sloan and Marshall, and they agreed that it is certainly a feasible idea. They said of course that they had given it much consideration, and they pointed out the Sloan tube is indeed a special case of such an idea, i.e., in which there is but one section—the first resonator. . . . It seems to me, however, that a matter of this sort is not susceptible to complete paper analysis, and I would be in favor of someone's experimentally developing the multiple resonator arrangement to find out its good and bad points. You know as well as I do that experimental work always brings to light things of which one does not think by any amount of cerebration. . . .

Back at Tuxedo Park, Loomis, having anticipated events, had a running start on the NDRC in more ways than one. His independent investigation of microwave radar was well under way, and as he wrote Lawrence, "We have a big group now going at Tuxedo on microwaves, it means that I am going pretty hard seven days a week." By the time various members of the microwave committee and the NDRC started beating a path to Tower House late that summer and early fall to see what Loomis' band of researchers had developed, an experimental apparatus had been constructed in the laboratory and was being tested with a makeshift moving Doppler target that consisted of a row of copper wires on a moving belt. The Doppler target could be run in the lab or set up in a nearby wood and detected by the lab set. Visitors were impressed that it was possible to tell whether the belt machine was running or not when it was remote from the actual detection device.

A second detection device had been assembled and installed in a large delivery van that Loomis bought expressly for that purpose. The staff immediately dubbed it "the didey wagon" because it had been used for delivering diapers. Loomis arranged to have the truck painted the traditional Tuxedo Park colors, green with gold trim, with "Loomis

Laboratories" handsomely lettered on the side. Whether this was to impress the dignitaries from Washington or to make the truck less of an eyesore to the neighboring swells, it lent the backyard enterprise a nice official air.

"When we took [it] out to the golf course, the thing was a microwave radar," recalled Lewis. "It had an indicator system that was capable of showing what we were looking at and getting us a reading on the speed. We took it out there and pointed it down the highway that went through Tuxedo Park. After we'd spent five minutes looking at it, I said, 'Hey, don't let the cops get a hold of that. These guys are all going over the speed limit.' Nobody else had one," added Lewis, who had immediately recognized the obvious application for their radar gun. "There weren't any microwave speed measuring sets in those days until we got that one."

The apparatus they had invented used an 8.6 cm klystron transmitter feeding energy into a novel antenna system called a "Tuxedo horn," which was a modified version of Hansen's leaky-pipe antenna. According to Lewis, after a considerable amount of work, they had settled on two so-called leaky pipes along two sides of a triangular horn, to make the radiator mechanically more solid and improve the shape of the beam. The Tuxedo horns were mounted together on a rotating platform, resembling a gun mount, protruding from the back of the truck. The receiver consisted of a crystal detector and audio amplifier. A "wave trap" fitted along the front edge of the two horns prevented the transmitted signal from leaking back into the receiver. According to the laboratory notes, the system's operation depended on the interference effects between the radiation directly picked up by the transmitter and the reflected energy of the moving target:

A moving object will produce a doppler effect, the signal returning from the moving object beating against outgoing signal, part of which is picked up directly by the receiving horn. The audio note thus received will be proportional to the radial speed of the moving object. Thus, a plane moving at a radial speed of 100 meters per second, thus producing a note of 2,000 cycles per second (assuming half wave length equals five centimeters). . . .

All that summer, Loomis and his team experimented on a variety of moving targets, using whatever was at hand in the sleepy resort. Hobart

tried testing the device on balloons he sent floating up over the tree-
tops. Outdoor tests were also done on motorboats in the lake, marked by
a corner reflector. As their experiments tracking automobiles progressed, they were able to measure the speed of the cars with "considerable accuracy: the Doppler shift being at the rate of 10 cycles per mile an
hour." In late August, they drove the "didey wagon" down to Bendix
Airport, where, according to Lewis, they managed "to follow without
difficulty the comings and goings of a number of small planes, [Piper]
Cubs and the like." A Luscombe, with its large metal surface, proved to
be the best target. Henry Loomis, now twenty-one, often piloted the
small Cub, which they learned they could follow at a maximum distance of two miles. Beyond that, the "fourth-power law" began to defeat
their efforts and the signals were lost in the noise. At one point, they
even tried to track the Goodyear blimp, but it moved too slowly to be
detected with their system. Some efforts to do experiments on the
movements of the Nyack ferry also failed, though in making their measurements and modifying the set to improve their range, Lewis and his
co-workers came very close to the idea of pulse radar.

    While they did not know it at the time, their leader had just been in-
formed of its existence during the army maneuvers in Ogdensburg. As
Lewis would learn later, Loomis understood full well the superiority of
the pulsed radar he had been shown but was not permitted to begin
doing similar experiments at his own facility without clearance, so he
"had to bide his time with the Doppler system, just to show the principles." Stymied by the lack of a tube that could supply enough power, the
microwave committee had decided to write a report. "A sign," as one of
its members observed, "that we didn't know what to do next." But before the end of summer, Loomis' work was interrupted by the dramatic
arrival in Washington of the British Technical and Scientific Mission,
led by Sir Henry Tizard, an influential defense scientist. When he hurried off to the nation's capital to meet with the British, Loomis had no
idea his days in Tuxedo Park were drawing to a close.

Chapter 9

PRECIOUS CARGO

> But first will you let me introduce my guests to you when they are all together? Some scientists are like prima donnas, you know, and they may be getting pretty nervous.
>
> —WR, from *Brain Waves and Death*

BY the summer of 1940, Britain was teetering on the edge of despair. Germany had begun relentless air attacks on England as a prelude to invasion, and Tizard, who was chairman of the key scientific committee on air defense, realized that they would not be able to stand alone for long. Hitler occupied a large part of Europe, and it was only a matter of time before England was outmatched by Germany's productive power. Tizard foresaw, with greater clarity than either the politicians or the military leaders, that the war had become harnessed to technology and technical superiority. The ability to produce powerful new weapons was the key to victory.

Convinced that Britain simply could not win without the assistance of the United States in developing and building these new instruments of war, Tizard had been lobbying vigorously for an overall exchange of scientific information of military significance. His idea was not greeted with enthusiasm at first, and Robert Watson-Watt, Britain's premier radar authority, went on record saying that the United States had nothing to offer. It did not help that an earlier fact-finding trip, headed up by the respected Nobel laureate Archibald Vivian Hill with the intention of gathering information about American science, had been a failure. Of course, with no secrets to barter and nothing to sweeten the pot, Hill had not been able to induce the Americans to divulge anything of value. But after France fell, and Hitler's forces were encamped along the Channel coast, England's new prime minister looked more favorably on Tizard's plan.

In one of the great gambles of the war, Churchill decided to support the idea of a technical mission to America and personally undertook the negotiations with Roosevelt. Events necessitated that they move very quickly. By July, the arrangements were embodied in an aide-mémoire signed by President Roosevelt and British ambassador Lord Lothian. In early August, Tizard picked the six men who were to take part in the British Technical and Scientific Mission to the United States, informally known as the Tizard Mission: Brigadier F. C. Wallace (British Army), a distinguished officer who had been in charge of the antiaircraft defenses at Dunkirk and had been one of the last men to leave the beaches; Captain H. W. Faulkner (Royal Navy), who was just back from serving in the Atlantic during the Norwegian campaign; Group Captain F. L. Pierce (Royal Air Force), who had made the first of many bomb attacks on the German destroyers hiding in the Norwegian fjords; John Cockcroft, the respected Cambridge physicist, who had built one of the first proton accelerators and at the outbreak of war became head of army research; Edward "Taffy" Bowen, a young Welsh physicist who was one of England's radar pioneers; and Arthur Woodward-Nutt, an Air Ministry official, who would serve as secretary. The plan was for Tizard to go ahead by air to Canada, which he had insisted be treated as an equal partner in the negotiations, and brief the Canadian government on what was to be disclosed to the Americans. He would then proceed to Washington to prepare the way. The rest of the mission would follow by ship.

Upon arriving in Washington on August 22, Tizard immediately made contact with Lord Lothian at the British embassy. He expected that A. V. Hill would have at least made administrative arrangements for the mission, but much to his dismay nothing had been done. "No office, no typists, etc.," he complained to his diary. "A good number of people do not know what I am here for!" As the embassy was too short on space to accommodate their needs, he hastily set up temporary headquarters at the nearby Shoreham Hotel, overlooking Rock Creek Park. Time was of the essence. Two days later, on August 24, the Luftwaffe executed a major strike against the Manston airfield and badly damaged the northeast London suburbs.

On August 26, Lothian arranged for Tizard to have an official audience with President Roosevelt. Because of the secrecy surrounding the mission, they slipped into the White House by the back door to avoid the press. Roosevelt welcomed them but seemed to be in a somewhat pessimistic frame of mind. He told them he was going to get his draft bill for conscription through Congress, but "it would probably lose him the election in November." He talked in generalities, except for explaining briefly that he was withholding the Norden bomb sight, one of the most important military secrets of the war. The device, which included an automatic pilot, used a mechanical analog computer to determine the exact moment a bomb should be dropped to accurately hit its target. It was then thought to be the key to daylight strategic bombing. The president told Tizard the reasons for not revealing it were "largely political," and if he could get evidence that the Germans had it, he could release it. Tizard then had a long meeting with the president's new secretary of war, who received him cordially.

"He seemed a very nice and sensible British scientist," Stimson observed of the Oxford-trained chemist, who had a pince-nez and the elegant manner that went with it. Stimson told Tizard to see Bush at the NDRC and recommended he talk to Compton and Loomis about radar, though he shrewdly noted that he already seemed "very well acquainted" with the scientific work being done in America. But the veteran diplomat knew that Tizard had made the trip across the Atlantic with a far more bold proposition: "He comes over here on behalf of his Government to offer us all of their secrets and hopes we may offer them some of ours."

Back in London, the six members of the secret Tizard Mission were making frantic preparations. Tizard had left them in no doubt "about the importance of the mission and the seriousness with which it was regarded" by Churchill. Their purpose, under painstakingly careful security procedures, was to hand over to the U.S. services all of their country's recent technical advances. That meant virtually every British secret: the jet engine, still in embryonic form, new antisubmarine devices, predictors, proximity fuses, explosives, and radar, in all its forms. In the hectic weeks before their departure, they rushed around collecting documentation on all the classified wartime developments: books, manuals, circuit diagrams, blueprints, films—anything that provided factual evidence of work in progress. Cockcroft set about collecting items of classified military equipment, while Taffy Bowen, the radar specialist, gathered together all his notes on their prized RDF system. Most of it would be packed in a large black metal deed box, of the kind ordinarily found in a solicitor's office, which Cockcroft had bought at an army and navy store. The box was kept under close guard in the headquarters of the Department of Supply at Savoy House in London, where Cockcroft kept an office, until their departure.

By far the most valuable item the mission would be bringing with it was a sample of one of the first resonant cavity magnetrons, a powerful source of microwaves that had been invented only seven months earlier at Birmingham University. From the moment John Randall and Henry Boot had conceived of the idea, to the dramatic results achieved with the first unfinished laboratory model, it was clear they had made a gigantic breakthrough in radar. No bigger than the pendulum of a grandfather clock, the copper disk was capable of generating high power (ten kilowatts) and very short wavelength (ten centimeters) radio waves. Nothing like it had ever been heard of, and most scientists believed such a device would not be within reach for years. Many decades later, Bowen could still recall "the drama of the occasion" as the news sank in—"the performance of the resonant magnetron was simply revolutionary." It would clearly lead to a new generation of compact, high-resolution radar that could breathe new life into England's beleaguered defenses and turn its planes into deadly night fighters and submarine patrollers that could operate under cover of darkness. It would also lead to the development of more accurate antiaircraft guns, which the British urgently needed to combat the dive bombers Germany was employing so effectively in advance of its armored columns.

For Britain, the magnetron was truly "a pearl beyond price." It was a technical miracle that could change the course of the war. The improved range and resolution of microwave radar could be a decisive factor in securing victory. All that was required was exploiting the new technology in a timely fashion. But in those last weeks of summer, the Luftwaffe had switched to night attacks, and sporadic bombing of London had begun. With its technical and industrial resources taxed to the breaking point, it was simply not possible for England to develop and mass-produce the new devices fast enough to make a difference in this war. Only with America's help could they capitalize on this stroke of luck. Threatened with invasion, the British could not afford to have this discovery fall into the hands of the Germans. They had to give the magnetron to the Americans if they wanted them as partners—it would be their dowry in marriage. It was a matter of survival.

The British also knew that many American scientists were eager to help England in its crisis. Sir Mark Lawrence Oliphant, the esteemed Australian physicist who was head of the Physics Department at Birmingham University, had received a letter from Lawrence earlier that spring, offering to give them the benefit of the United States' advances in radar: "There has been a good deal of progress in this country on microwaves, and I do not know why Dr. Hill has not been able to get the information you want. I think it has to do with the commercial aspects," Lawrence wrote, referring to the complex government restrictions on the interchange of patented information. "I have given him the best advice I could." Loomis had again broached the subject of cooperation in these crucial times during his visit later that summer.

In those final weeks, Bowen made a special trip to the General Electric Company's research laboratory in the London suburb of Wembley for a detailed briefing on the new resonant magnetron. The first twelve production models had just been completed a few weeks earlier, and after they were run through a test rig, he selected the best one to take with him to the United States. As it happened, he chose number twelve, which would later turn out to be significant. A few days later, he returned to pick it up. Sticking to his usual mode of transportation to attract as little attention as possible, he carried it to London by underground and with great relief deposited it in the tin trunk, now filled to the brim with Britain's most valuable military secrets.

Though not yet thirty, Bowen was the leading defense scientist on the mission and an expert in all of the country's top-secret radar systems. He was to be "custodian of the black box" on the journey across the ocean. Bowen, with the black box, was to travel separately to the dock at Liverpool, and he was expected to find his own way to the ship in which he and the rest of the team would be making the crossing. So late on the evening of August 28, he showed up at the back door of Savoy House, where he had arranged for a guard to hand off the bulky package. Bowen took it by taxi to the Cumberland Hotel, where he had planned to store it overnight in the safe. But when the hotel manager saw the box, he shook his head in dismay—it was larger than the safe.

Bowen had no choice but to stow the kingdom's treasure chest under the bed in his locked hotel room and count the hours until morning, when he was to catch the eight-thirty A.M. boat train for Liverpool from Euston Station.

Early the next morning, Bowen hailed a taxi to take him to the station. The driver stubbornly refused to let him keep the box on the seat next to him and instead insisted on strapping it to the roof. Bowen could not afford to waste time arguing, and, as he later told the story, "We made the short run to Euston with that supremely important piece of luggage prominently displayed on the roof." At the station, things went from bad to worse, and he almost lost England's last best hope in the rush-hour stampede:

With my luggage, the box was more than I could handle, so I called a porter and told him to head for the Liverpool train. He grabbed the box, put it on his shoulder and headed off so fast that (an old cross-country runner and still pretty fit) I had great difficulty keeping up with him. He got well ahead and the only way of keeping track of him was to watch the box weaving its way through the mass of heads in front. A first class seat had been reserved for me, but beyond that I did not know what to expect. All of the other members of the Mission were going to Liverpool by different routes, and I was alone on this leg of the journey.

I found my seat and, with the black box parked on the luggage rack, waited for departure time. What I had not realised was that the whole compartment had been reserved and when I entered it all the blinds were drawn and large notices were posted on both windows. A few minutes before departure time, an exceptionally trim, well dressed gentleman with a public-school tie came into the compartment and with scarcely a glance in my direction settled down in a seat diagonally opposite reading The Times. Not long after the train started, the late arrivals started shuffling up and down the corridor looking for an empty compartment. A couple of bright sparks opened the corridor door and said, "Here we are, chaps, this one is nearly empty," and started to enter. My companion spoke up for the first time and said, "Out. Don't you see this is specially reserved?" It was not what he said but how he said it. The would-be intruders wilted and we had no further interruptions. For the first time, I realized the precious cargo was under some kind of protection.

Bowen's instructions were to stay put in his compartment until the box was picked up by the army, so when the train came to a stop in Liverpool he remained seated. He noticed that his silent companion also showed no signs of leaving and appeared completely absorbed in his *Times*. Bowen gradually became aware of the sound of marching feet, and soon a squad of a dozen fully armed soldiers materialized on the platform. Led by a sergeant, they executed a series of complicated maneuvers and, "with much slapping of rifle butts," came to a halt alongside his compartment. On a barked command from the sergeant, they stood at ease, and Bowen watched as one member of the squad stepped forward and opened the compartment door: "Another two collected the black box and trundled it outside. On a further word of command, they shouldered it and marched off in the direction of the ship." Throughout the impressive performance, his dapper companion had "not moved a muscle." The moment the squad left, however, "he rolled up his newspaper and, with a barely perceptible nod in my direction, took his departure into the corridor."

The presence of such a large military escort was quite comforting, and the young physicist was just starting to relax when a terrible thought crossed his mind: "I was beginning to feel that things were well looked after," he would later recall. "Alternatively, if this was the enemy making off with Britain's secrets, they were making a spectacular job of it."

Bowen hurriedly boarded the ship, the *Duchess of Richmond*, and ascertained that the black box had indeed been delivered and was under guard on the bridge. He was informed that the captain's instructions were that "in the event of an enemy attack and a likelihood of the vessel being lost, the box was to be heaved over the side and allowed to sink in the mid-Atlantic." Only the group secretary, Woodward-Nutt, would be allowed access to the black box during the voyage. It was arranged that he would meet a third officer, who kept the keys to the locked strong room, should they need to dump their secret cargo into the drink.

They set sail that evening as darkness fell and headed for the Irish Sea. They had not gone far when bombs started to drop from the sky, falling all around them. Nighttime bombings were not uncommon, and it was probably part of an air raid on Liverpool. Fortunately, the Germans were not lucky that night, and none of the stray warheads struck the ship. But it was enough to stop them for the night, and they dropped anchor. The next morning they set off again, flanked by boats that served as minesweepers for the hazardous journey down the Mersey

River. The *Duchess* was an unescorted Canadian passenger liner and relied on speed to make the crossing safely. To elude German U-boats, she made regular course changes every twenty to thirty minutes. It made for a rough passage, Bowen recalled, and the boat earned the nickname the "Drunken Duchess" for the way she would "roll all over the high seas."

The six members of the mission were accompanied on their journey by a thousand British sailors, brought over to man the first of the fifty overage destroyers that the American government was providing in exchange for the use of Britain's naval bases in the West Indies. (Known as the "destroyers for bases" deal, it laid the seeds for the lend-lease bill, which Roosevelt proposed in response to Churchill's desperate request that December that America provide Britain with the guns, tanks, and ships "to finish the job" and defeat Germany.) Rumor had spread among all the navy men that a famous physicist was on board, and they asked if Cockcroft could find the time to give a lecture during the week's voyage. As he could scarcely talk about his assignment or any of the contents of the black box, Cockcroft searched for a safe topic with which to entertain the bored servicemen. His choice, ironically, was atomic energy, still considered years away from being realized and of no possible importance to the war. He greatly impressed his audience, however, when he informed them that the potential amount of atomic energy in a cup of water could blow a fifty-thousand-ton battleship one foot out of the sea.

The rest of the trip was uneventful, though Bowen would always remember a conversation he had with Cockcroft in the bar one evening before dinner as an example of the kind of calculation only a physicist would make. Cockcroft had been worrying about their secret cargo and posed the question to Bowen that if the ship were indeed attacked, and the trunk thrown overboard, would it sink or swim? Cockcroft's conclusion was "open and shut": the black box and all it contained would surely float. Bowen gave the matter no more thought until he saw the box again when they sailed into Halifax harbor in Nova Scotia on September 6. Waiting on shore was an armored vehicle loaded with submachine guns, accompanied by an even bigger armed guard than the one that met the train, to transport the black box to Washington. As the box was being moved, Bowen's eyes widened in surprise—"a neat pattern of holes had been drilled in each end." Cockcroft had apparently seen to it that the crate would sink.

Woodward-Nutt turned over the secret cargo to the Canadian military guard to transport to the border, where it was to be given to American authorities and taken directly to the British embassy in Washington. But upon his arrival at the embassy a few days later, he "was a bit shaken to find that the samples and documents that I had seen so carefully off at Halifax had not yet arrived." After a series of frantic phone calls, the missing bounty was located and sent along its way. When it finally arrived on September 9, the black box was stashed in the embassy's wine cellar, and the sole key to the door was entrusted to the ambassador's butler.

B Y the time the six members of the mission were united in Washington on September 11, Tizard had made all the preliminary arrangements, and they were scheduled to officially hand over the British secrets the following day. The first and most important matter on the agenda was radar. The next two or three weeks would be taken up with a series of meetings designed to acquaint the top levels of the U.S. Army and Navy with all the new British developments and how they were being used. Then it would be the Americans' turn to show the radar systems it had developed up to that time. This would include visits to the Naval Research Laboratory at Anacostia, in Washington, D.C., and the Signal Corps Laboratory at Fort Monmouth, New Jersey. Although there were startling similarities between some of their radar systems, it quickly became clear to all concerned that the British were way ahead.

Working at a faster pace than the United States because of the immediate threat to their security, the British had developed a system of radio detection by a pulse method, and the first experimental system had been set up on a small island off the east coast of England. Between the fall of 1935 and 1938, a highly efficient system of radar stations, called the Chain Home network system, was erected to give advance warning of approaching aircraft. By September 1940, the Chain Home system had been in operation against the enemy for over a year and was at that very moment helping the British pick off German planes as they streamed over the English Channel. Combined with their quick-response fighter units, the British were tearing up Hitler's forces.

The U.S. Joint Chiefs of Staff were floored by the reports that the air warning system was proving a decisive factor in the Battle of Britain, from which England was already emerging as a victor. When the air assault had begun earlier that summer, the Germans had been repelled with heavy losses, in no small part because the British were forewarned of each assault by their radar detectors and could order their Spitfire and Hurricane pilots into the most advantageous position to intercept them. It was impossible for the American military leaders not to be deeply affected by the facts laid out before them—radar was emerging to all the world as a vitally important weapon of war. As a direct result of those first meetings, the top army and navy brass took immediate steps to bring their long-range radar (at the meter wavelength) up to speed and introduced into service.

The British had not revealed their most valuable secret, the existence of the magnetron. It had been hinted at, and the possibility of generating substantial pulse power at short centimeter wavelengths had been suggested, but the mysterious device that could achieve this had not yet been described. When Bush and Tizard had first met on August 28 at the Cosmos Club, the highbrow Lafayette Square meeting place, Bush recalled the delicate tap dance they both kept up, as there were still many bureaucratic stumbling blocks to free discussion. While waiting for the army and navy to obtain official authorization for the NDRC to talk to the British, Bush arranged frequent meetings with his British counterpart "behind the barn":

> We were rather careful not to seen together too much for fear that some people in Washington would conclude that there was some sort of conspiracy under way. Both Tizard and I were in something of a quandary, for each of us would have liked to have told our own group that the other had made very significant advances which would be valuable, but we could not do that as we were both of us under severe limitations. I think you can well imagine the skill with which Tizard handled the affair, making contacts with dozens of people and giving the impression, without saying anything definite whatever, that Britain had made an extraordinary advance. . . .

It was only natural that the American military officials were, at the start, doubtful as to whether the British were "putting all our cards on the table," as Cockcroft put it. There was a deep-seated conviction of technical superiority on both sides and a shared suspicion that the other was withholding some secrets. Cockcroft used his long-standing friendship with Loomis and Lawrence to try to establish an atmosphere of greater trust and warmth. By special arrangement, Cockcroft and the members of the Tizard Mission were authorized to brief civilian scientists like Loomis on the same basis as top military officers.

On the evening of September 11, Loomis invited the Tizard Mission to an informal evening conference at the Wardman Park Hotel, a short walk from their headquarters at the Shoreham. Loomis began by outlining everything his panel had been doing in the microwave field. He filled them in on the recent tour of industrial laboratories working on microwaves he and Compton had conducted that summer, detailed which research facilities they would be advised to visit, and invited them to Tuxedo Park to see what was being done in his own laboratory. He also explained the progress that had been made in transmitting and receiving tubes at wavelengths around forty to fifty centimeters, which outstripped what they were doing in Britain. But in the very short wavelengths, they had yet to find their way past the same obstacle that had once stalled British physicists—namely, the lack of a high-power oscillator.

When the British expressed their interest in microwave radar, Loomis reacted with great enthusiasm. The standard radar sets used by the United States Army and Navy all operated on the same relatively long wavelength of one to two meters, with correspondingly broad beams. Similarly, the Chain Home stations were large installations with the massive reflectors needed to accommodate the ten-meter wavelengths the system used. Bowen, who had spent the previous five years setting up those reflectors in some of the most barren, godforsaken spots on earth, appreciated the American's recognition of what they had accomplished. At the same time, Loomis grasped the considerable military advantage that could be had in developing the still primitive radar technology in the very short "microwave," or ten-centimeter wavelength, range, which would allow them to gauge the enemy position more accurately and distinguish among tightly bunched targets. The young physicist took an immediate liking to Loomis—he was unquestionably an original and stood apart from the others in the room. He was also openly and resolutely in favor of helping Britain win the war, which permanently endeared him to the Welshman.

After the dinner, they retired to the large suite Loomis kept at the Wardman Park for just such occasions. The American military officers were surprisingly reticent. The attitude of one admiral, in particular, struck Compton as very "anti-British," and he appeared to spend a good part of the evening becoming conspicuously drunk, which Compton suspected was "a dodge" to avoid giving Tizard any more information than necessary. Loomis, on the other hand, established an easy rapport with Tizard, whose background in many ways mirrored his own. They both belonged to their country's elite and possessed the skilled, patient, genial manner that disguised a razor edge. Both had worked in research and development during the previous war; and what Loomis had done for gunnery at Aberdeen, Tizard had accomplished in improving the technology of British aircraft. In a room filled with stuffy admirals and generals from both sides of the ocean, the physicists felt a strong sense of camaraderie, bolstered by their mutual confidence and admiration. The nucleus of a powerful alliance had been established.

Before the night was out, despite what Cockcroft felt to be a "rather doubtful opening," the British quietly revealed the magnetron to the small party of Americans in Loomis' suite. Looking from face to face, Bowen could see they were stunned to learn that it could produce a full ten kilowatts of pulsed power at a wavelength of ten centimeters. "The disclosure was the key point," recalled Cockcroft. "From then on we had no difficulties." The decision was quickly made that an American team should start at once on the development and production of ten-centimeter radar based on the British magnetron, for use both on the ground and in the air. The army and navy opted to proceed with the development of their own longer-wave radar systems, and Loomis and his newly formed microwave committee took responsibility for pursuing the experimental short-centimeter waves. In the days that followed, Tizard formalized these arrangements with Bush, and Bowen and Cockcroft made preparations for the crucial meeting with the microwave committee to be held that weekend at Loomis' place in Tuxedo Park.

On the morning of Saturday, September 28, Bowen and Cockcroft flew to La Guardia Airport in New York City, where they were met by one of Loomis' chauffeured cars and ferried in grand style forty miles to Tuxedo Park. The leaves were at the height of their autumn glory, and the two British scientists admired the colorful display bursting from the elms, oaks, and maples lining the long, twisting road that led up through the hills to the mansion-cum-laboratory. Loomis greeted them and took them on a tour of Tower House and a second structure he called his "playhouse," which Bowen recalled as an ultramodern building with full air-conditioning and "automatic everything else." That afternoon, they all drove to Bendix Airport, and before a group consisting of Bowen,

Cockcroft, the MIT radar specialist Ed Bowles, and Carroll Wilson, Bush's executive secretary, along with several military and naval officers, Loomis staged a field demonstration of his ten-centimeter Doppler radar system, which easily detected a small two-seater plane at a range of two miles. While several of those present had never seen a working radar set before, Loomis was already aware that his device had been rendered obsolete by the magnetron. Afterward, at a meeting at the Tower House, he informed them about the microwave work being done at Berkeley and explained the possible uses of radio yttrium for "intelligence purposes," outlining the use of a small portable Geiger counter that he had helped develop. The British were encouraged by what they had seen, and Bowen cheerfully recorded in his notes at the end of the day that "Dr. Loomis" had organized "a powerful and energetic group, and it should be of great potential help to us."

The following evening, the two British scientists finally unveiled their prize, the secret invention they had kept successfully from the Germans and had shepherded across the ocean to bestow on the Americans. In Loomis' elegantly appointed sitting room, Bowen and Cockcroft produced a rough wooden box and unfastened the thumb screws that secured the lid. "The atmosphere was electric," Bowen recalled, as he slipped out the coppery disk that fit neatly in the palm of his hand. "They found it hard to believe that such a small device could produce so much power and that what lay on the table in front of us might prove to be the salvation of the Allied cause."

Cockcroft brought out the blueprints, complete with drawings and construction details, and spread them out on the floor. During the conversation that started that night and continued the next morning, Bowen and Cockcroft explained how the magnetron "opened up new vistas" and "increased the power available to U.S. technicians by a factor of 1,000." They also explained their objectives in radar, specifically the critical importance to their defenses of microwave airborne radar sets—the one piece of equipment, above all others, that could transform their defense against the night bomber. It would also enable the RAF's bombers to navigate the dense cloud cover that often obscured German fighter planes right under their noses and allow them to target factories and forces on the ground with far more accuracy. British planes equipped with extremely short centimeter radar would be able to detect U-boat periscopes piercing the fog, which was like picking a needle from a haystack for the longer-wave systems. Ed Bowles could scarcely believe what he was hearing: "All we could do was sit in admiration and gasp."

Loomis' excitement mounted as the magnetron's mode of operation and capabilities were explained. With a surge of excitement, he realized that here was the opening he had been looking for. The moment the British put the magnetron on the table, he knew "a major breach had been made." The klystron, which had been their best bet, could never be a high-powered oscillator. "This was the first fortunate break, for the magnetron itself had been hit upon by accident by Oliphant and his men." It changed everything. They now had a technical advantage over the Germans that they had to exploit immediately. As a veteran sailor, Loomis knew how important the element of luck was in deciding the outcome of any race: "The boat ahead gets the new breeze first, just because it is ahead, and thereby increases its lead." And this was one lead they could not afford to squander.

According to Cockcroft, right then and there Loomis proposed the idea of establishing a large central microwave laboratory. The British enthusiastically seconded the idea, and it was quickly agreed that it should be a civilian rather than military operation, staffed by scientists and engineers from both universities and industry, based on the British model of successful research laboratories and, not coincidentally, Loomis' own enterprise. Loomis suggested that the laboratory be based in Washington and run under Bush's auspices at the Carnegie Institution's Department of Terrestrial Magnetism. The new lab should begin work at once on the development of new ten-centimeter radar systems based on the magnetron.

What had begun as an introductory meeting the night before had by early morning evolved into a strategy session. Loomis called for a luncheon meeting to be held the very next day in New York, on Monday, October 1, to be attended by those members of the microwave committee who were not present. The magnetron was the break Loomis had been waiting for, and he was determined to make the most of it, even if it meant using all the reserves of American industry and science. It became, at once, the focus of Loomis' whole being. According to Bowen, they then spent hours burning up the phone lines as dozens of people were told what had transpired and where to report the next day: "I have no doubt that the New York Telephone system did an unusual amount of business that night because by the time Cockcroft and I left Tuxedo Park the next morning the Bell Telephone Company had been told that if the Boot and Randall magnetron came up to expectations, they would be given a contract by the Microwave Committee for a small production batch."

The two British physicists were overjoyed. Up to then, only twelve models of the magnetron had ever been made. Now it was to be put into production and could make a critical difference in almost every theater of the war—on land, in the water, and, most important, in the air. Finally their RAF fighters would be given a way to combat their chief menace, the night bomber. When Tizard was informed of what had been decided at Tuxedo Park, he was delighted. "Push it for all you are worth," he told Bowen on October 2 before flying out of New York by Pan Am Clipper, which would take him to Bermuda and then via the Azores to London. In spite of the tight security surrounding the mission, he was caught by photographers at the airport.

Later that day, Loomis called on Stimson in Washington to brief him on the dramatic new developments in radar:

> Alfred Loomis came in in the afternoon, full of excitement over his interviews with the British and with the scientists, and he was full of the benefits that we were getting out of the frank disclosure by the British to us of their inventions and discoveries of methods they have made since the war. He said we were getting the chance to start now two years ahead of where we were. . . .

The secretary of war was impressed by what he heard. He had already been alerted to radar's importance in the Battle of Britain by his cousin, but now he called in General Marshall to hear Loomis' view of the situation:

> Alfred was very flattering to the Army on the comparative frankness which we had shown on our part, as compared with the tendency on the part of the Navy to hold back all their secrets, even though the secrets did not amount to much. He said that our frankness had enabled the British to put their hands on the table and had given us a rich mine of information and it has also brought the Navy around to a better state of frankness on their part. Apparently everything is going along well now.

That night, Loomis joined Stimson for dinner at Woodley, the stately eighteenth-century mansion that was only a few minutes' drive from the old State, War, and Navy Building, next door to the White House, where Stimson kept his office. Several men from the War Department were also present, as were Brigadier Charles Lindemann, an attaché at the British embassy, and John J. McCloy, who had recently left the law firm of Cravath to become one of Stimson's top aides. McCloy, like most members of Stimson's inner circle—including First Assistant Secretary Robert Patterson, Assistant Secretary of War Robert Lovett, special assistant George Harrison, and Loomis—was a fellow graduate of Harvard Law School, who had been successful on Wall Street before coming to the War Department. They were all Republicans serving a Democratic president, who, apart from his party affiliation, was very much of the same stripe—the same class, social background, and education. They worked extremely closely together as a team, bound by strong ties of family, friendship, corporate and institutional loyalties, and, above all, their loyalty to Secretary Stimson and the ideals he was fighting for. "We had a very pleasant dinner," Stimson wrote in his diary of the tight-knit group, which gathered weekly in different numbers at his home, and "afterwards sat around talking, mainly over war problems, until eleven o'clock."

LOOMIS' microwave committee immediately placed a production order with the Bell Telephone Company for an exact copy of the magnetron. On October 3, the day after Tizard's departure, Bowen paid a visit to Bell's headquarters at 463 West Street in New York. A group of Bell's top researchers were waiting for him and were fairly bursting with curiosity to see the British magnetron, which they had heard about from Loomis. Bowen put the magnetron on the table and took out drawings of the internal construction and manufacturing details. "They were fascinated by the design," and if Bowen sensed that they had "reservations about the extraordinary output" the British had boasted of, they soon conceded it had "enormous potential." Not only did Bell's engineers anticipate no difficulty in copying the magnetron, they wanted to begin right away. If the device lived up to its promise, they would be awarded the plum manufacturing contract. Since it was advisable to first conduct a test run, arrangements were made to meet the following Sunday at Bell's Whippany laboratories in New Jersey, where they had a pulse generator and the requisite voltage.

Bowen left the magnetron in their care, and he and Cockcroft kept to a grueling schedule, visiting the laboratories Loomis had recommended, beginning with RCA and MIT, and later General Electric, Western Electric, Sperry, and Westinghouse, among others. At each stop they doled out just enough technical information to leave their audience intrigued and were careful to hold back information on the magnetron, which they were saving for Bell. On Sunday, October 6, Bowen drove to the Whippany laboratories, where several of the senior Bell staff were gathered. As Bowen surveyed the equipment they had dug up, including a large electromagnet of the sort common in university physics departments, he was painfully aware of the fact that the last time the magnetron had been tested was two months earlier back in Wembley. It had traveled three thousand miles, by train, boat, and armored vehicle. "Very gingerly," he recalled, "we switched on the anode potential and were instantly rewarded with a glow discharge about an inch long coming from the output terminal." Bowen stared at it, blinking in surprise: "I had not seen anything like this before and we looked at it in amazement."

One of the Bell engineers asked, "And what wavelength do you think that is?" Bowen's off-the-cuff estimate put it at around ten centimeters. "And how much power do you think it is giving?" Bowen had been quoting the figure he had been given of ten kilowatts, but he knew the glow discharge he was looking must be higher. In fact the magnetron had generated a pulse power in excess of fifteen kilowatts at a wavelength of 9.8 centimeters—"as a demonstration it could not have been more successful." Bowen then learned that Loomis had already swung into action and had told Bell that if the test run went as planned, they were to immediately commence manufacturing thirty copies. When Bowen departed, he was "walking on air."

But by the law of averages, Bowen should have known that everything had gone too smoothly. The next day brought a terse call from the director of Bell Labs, Mervin Kelly, who had worked himself up into quite a state. Bowen's presence was urgently required in New York— "right now or tomorrow morning at the latest." While Kelly could not spell out the problem on an open phone line, clearly something had gone very wrong. "Oh, my God," Bowen thought, "the magnetron must have blown up." He did not have a clue what the problem was and spent an uneasy night wondering if the whole mission had foundered only days away from the magnetron's being put into mass production. He caught a six forty-five A.M. flight to New York and went straight to Bell's headquarters on West Street, where he was ushered into the top-floor conference room. There, Kelly, acting like judge and jury rolled into one, and flanked by two or three others who "looked suspiciously like company lawyers," produced the magnetron and laid it on the table. "Is that the magnetron you brought here last week?" he demanded. Bowen, still mystified, had to agree that it was.

There was a long pause, and then Kelly produced a photograph. It seemed that in an effort to gain more information about the magnetron, Bell's engineers had taken an X-ray picture of the device. As Bowen put it, "The result came as a considerable shock." The magnetron, which had demonstrably worked very well, had eight holes, not the six clearly specified in all the blueprints and manufacturing material supplied by Wembley. Bowen looked from the magnetron to the photograph. There was nothing he could say. The unspoken accusation that the British had pulled a fast one hung in the air.

Bowen asked if he could make a transatlantic call to clear up the matter, and in moments the Bell people had Eric Megaw, the resident genius of GE's Wembley plant, on the line. At first Megaw was as confused as Bowen and at a loss for an answer. Then "the penny dropped." He remembered that on the afternoon he had placed the order for the first twelve magnetrons, he had gone back at ten P.M. that same night and had told the foreman, "Look, on that order for twelve magnetrons, make ten with six holes, one with seven, and one with eight." Later, rushed and under tremendous pressure, he had completely forgotten about it. When Bowen had gone to Wembley and picked out the best magnetron to bring to America, he had selected the last model on the line—number twelve—the one with eight holes. As soon as the discrepancy was explained, the air cleared. But now they were faced with a new problem: Should Bell build their copies based on the blueprints for the six-cavity magnetron supplied by the British or from their X-ray picture of the eight-hole device? "We agreed there was only one way to go," recalled Bowen, "to copy the one they knew worked." The Bell copy followed the X-rayed model to the letter, which is why all the early British magnetrons had six holes, while all the American versions had eight.

WITH Germany threatening to bomb Britain into submission and moving in submarines to blockade shipments of food and supplies, events demanded quick action. Britain was involved in a life-and-death struggle, as Cockcroft had made painfully clear. Loomis knew that getting the laboratory up and running, and new radar systems developed in time to rout Hitler's forces, was going to be a massive undertaking. It would take American ingenuity and American money. Before he left Tuxedo Park, Loomis put in a call to Ernest Lawrence. Since their success with the great cyclotron, they were regarded as something of a dynamic duo. Now the crisis in Europe required that they work their magic one more time and together create another astoundingly powerful device.

The sense of urgency that suddenly prevailed was expressed in the telegram Bush sent Lawrence summoning him to Washington:

EARLY TRIP DESIRABLE IMPORTANT MEETING BOSTON
TENTH STOP YOU SHOULD CONSULT LOOMIS AT
LENGTH BEFORE THEN STOP SECRETARY SENDING YOU
APPOINTMENT AND TRAVEL ORDERS AIRMAIL
ARRANGE DETAILED PLANS WITH LOOMIS AS YOU WILL
BE ATTACHED TO HIS SECTION.

Lawrence, who was still immersed in pure science, was not convinced he should make war work his first priority. He wanted to stay on the job in Berkeley and believed that was the best use of his time. Bush had prevailed on him on behalf of the NDRC once before, when a month earlier he had asked him to head a roving committee—"a sort of fire department"—to call in on any problem that might arise. Lawrence had made various excuses, and the impatient Bush had finally replied, "Forget the whole thing." But this time it was Loomis who was requesting his services. He sent telegrams to both men saying he was on his way and left for New York on October 5.

Lawrence took the oath of office required to work for the top-secret radar division of the NDRC, then Loomis brought him up-to-date on the secrets contained in the black box and all the plans that were under way. On Thursday, October 10, a morning meeting was held at Compton's office at MIT that lasted into late afternoon. The following day, Loomis held a meeting of the microwave committee at the Carnegie Institution on P Street in Washington, attended by Bush, Compton, Bowles, and Bowen. Loomis introduced Lawrence, the famous Nobel laureate, to the radar specialists. Their purpose was to establish a civilian research laboratory, as proposed by Loomis at Tuxedo

Park. It was decided that they would begin with a nucleus of about twelve scientists, to be recruited from universities, and then expand it as rapidly as possible. Each scientist they recruited would be asked to recommend the names of four or five other candidates, then they would be contacted, and so forth. Bush, who was worried about potential conflicts with Carnegie, rejected the idea of basing the lab at the Department of Terrestrial Magnetism. Bowen also pointed out the importance of having an experimental aircraft at their disposal. Loomis suggested they look into Bolling Field, a U.S. Army Air Corps base on the outskirts of Washington, which had a large heated hangar that he could have quickly fitted up as a laboratory for experimental work.

That weekend they adjourned to Tuxedo Park, where the talks would continue in a more relaxed setting. They were joined by Cockcroft and, in Bush's absence, Carroll Wilson. Bowen and Cockcroft once again pulled out the magnetron and blueprints and explained to Lawrence what the small device could mean. Lawrence had been very pessimistic about Britain's ability to survive the Blitz and felt nothing could save England in the face of that onslaught. But listening to Bowen and Cockcroft, he envisioned for the first time the many possibilities opened up by the magnetron and very short radio waves. "This is something you can get your teeth into!" Lawrence told them.

Loomis then laid out the three main projects that needed to be tackled first and paid tribute to the Tizard Mission by indicating that they were the radar systems most urgently needed by Britain. Their first priority would be to develop the magnetron into a ten-centimeter airborne intercept system, code-named AI-10. Second, they would work on high-accuracy ten-centimeter gun-laying (GL) radar for antiaircraft use. Third on Loomis' wish list was a long-range aircraft navigation system, of a yet unspecified type.

After a hearty English breakfast laid on by Loomis' expat butler, they discussed which areas America could be of most help in. Tizard was convinced that North America was the ideal place to work on a long-range navigational system, primarily because Britain was severely limited in its ability to test such a system without risk of being shot at by hostile aircraft. The morning's discussion was animated, but Loomis and his guests did not overly exert themselves. It was Saturday after all, and they had been hard at it for weeks now. Everyone was in need of a little rest and recreation. In the afternoon, Bowen and Lawrence played tennis, and the Welshman found himself being "thoroughly beaten" by the Berkeley physicist. Before dinner that night, a Tuxedo neighbor, Averell Harriman, dropped by to talk and to have a quiet drink. Like Loomis, Harriman had been drafted as one of Roosevelt's "dollar a year" wartime advisers. Loomis talked about his large motor-powered yacht, which he had just sold to the navy for "the bottom price of one dollar" to be refitted for service as a troop carrier. Bowen could not help noticing that it was "a very different world" from the one in which he had grown up in the Swansea Valley.

The next morning, the foundations of a great radar research laboratory were laid. Loomis, backed by Bush and Compton, wanted Lawrence to be the new laboratory's director. Lawrence did not think he could delay work on his 184-inch cyclotron for that long but promised to help with the initial planning and organization, as well as enlisting the very best scientific minds in the country. But that raised an interesting question: Where would so many qualified scientists come from? Cockcroft and Bowen described how academic scientists had been drafted into radar research in Britain and had taken to the military work "like ducks to water." Loomis asked Bowen to write out detailed specifications for an airborne radar system, while Cockcroft drafted the corresponding specs for a gun-laying radar. Loomis then threw open the subject of long-range navigation for discussion, and the British outlined a proposed system—which later became known as GEE—in which ground stations transmitted a grid of radio beams that enabled aircraft to fix their locations within the distance of a couple of hundred miles. It was generally agreed, recalled Bowen, that the range should be one thousand miles or more, with an accuracy of plus or minus five miles.

Bowles, whose early resentment of Loomis was hardening into dislike, watched as the former Wall Street financier milked the British physicists for all they were worth:

> He took a great liking to Taffy Bowen, realizing here was a source to be cultivated. Loomis' nature was such that he would have done very well as a G2 agent, it was very clear he was going to do his best to extract every vestige of information he could from Taffy Bowen on what the GEE navigation system comprised. He was the kind of person who if one didn't know him, could seem to be operating very innocently. It was clear to me what he was up to.

According to Bowles, after they had all dressed for dinner and gathered for drinks downstairs, Loomis came in and announced that he had just had an epiphany in the shower. After taking Bowen and himself "quietly out of reach of hearing of anyone else," Loomis told them "he suddenly knew what kind of system we should develop for navigation. Taffy Bowen looked at me and winked and I knew damn well that Loomis had extracted all the information he needed so we could proceed as if we had a disclosure from the British of what they had." Bowles refused to believe that Loomis could have come up with such an idea on his own and immediately jumped to the conclusion that Bowen must have been the source of his "inspiration." He was livid that Loomis had used his "devious charm" to get one step ahead of him and the rest of his American colleagues and was now in possession of top-secret details of the GEE system, which the British had not disclosed officially to the United States.

Before the weekend was over, Loomis proposed that since the magnetron was already being manufactured, the first priority of the new lab was to assemble a ten-centimeter radar system that could be tested against echoes from ground objects in the same manner as he had been doing in Tuxedo Park with short-wave systems. The British and American scientists then compiled a list of everything they would need, including desirable figures for things like the pulse width, receiver sensitivity, cathode ray tube display, and so forth. Grabbing notepads and scribbling on the backs of envelopes, "We sketched the block diagram of a typical system right there," recalled Bowen, with all of the components: a modulator, a transmitter based on the magnetron, a receiver, an indicator, and adequate power supplies. Much to their surprise, Loomis then announced he was convening a meeting of the microwave committee for the very next day to allocate the contracts for the components.

Early Monday morning, October 14, Loomis and company drove to New York City and gathered at his penthouse apartment at 21 East 79th Street at eleven A.M. Due to the short notice, not all the committee members could make it, and the representatives from both GE and Westinghouse were absent. In his blunt fashion, Loomis informed the industry representatives present of his intention to begin building the ten-centimeter air interception system. "This visibly shook those who had not been in Tuxedo Park," recalled Bowen. Loomis "further astonished" the committee members by asking each one to select those parts of the project he wished to tender bids to manufacture and to deliver his contribution in thirty days. The industry representatives from Sperry,

Bell, and RCA replied that they saw no difficulty in getting their bids in under a month's time. "No," Loomis told his stunned audience, "you misunderstood me. I want you to submit your tender next week and *deliver* thirty days after that."

For a moment, everyone in the room was speechless. Rather than risk Loomis' displeasure, the committee members decided to reserve judgment until they could get back to their respective companies and consult with their colleagues. But one representative complained, "This is a funny committee—they tell *us* how long to take to do things." Pressed for an answer, however, Mervin Kelly confirmed Bell's interest in manufacturing the magnetron and disclosed to the others for the first time that they were, in fact, already under way. This revelation spurred the others into action: GE asked to make the magnet; RCA, the pulse modulator, cathode ray tubes, and power supply; Sperry, the parabolic reflectors and scanning gear. They all put their dibs in on the receivers and indicators. Loomis later negotiated with GE to supply the magnets and contracted Westinghouse to deliver the antennas. Bendix, an aviation company, won the bid to provide the power supply, which would ultimately run off an aircraft engine. The British scientists marveled at the dispatch with which the New York banker made decisions and handed down orders: "Only five weeks had passed since the arrival of the Mission in Halifax," Bowen reflected, "and things were moving exceptionally fast!"

After the meeting broke up, Loomis asked Bowen and Lawrence to stay behind. He then turned to Lawrence: Would he be willing to act as the principal recruiting agent for the new laboratory? Lawrence not only agreed, he immediately volunteered two of his very best students, Luis Alvarez and Edwin McMillan, though he was canny enough to suspect that his old friend had already had his eye on both of them. Lawrence had done a little advance work of his own before coming to Tuxedo Park. On October 8, right after being briefed by Loomis, he had gone to Harvard to meet with Kenneth T. Bainbridge, the respected physicist who had built the university's cyclotron. As they walked the streets of Cambridge, Lawrence, swearing his colleague to silence, described the radar project and asked if he was prepared to undertake "some important secret research."

Lawrence, whom Bowen observed was a man of sudden enthusiasms, "began using the telephone right there in Loomis' apartment, and the telephone bill must have been astronomical." The following day, on

October 15, Lawrence, still operating out of Loomis' sitting room, called up his friend and protégé Lee DuBridge, chairman of the Physics Department of the University of Rochester and a perfect candidate for the new laboratory's director. At thirty-nine, DuBridge was popular and well regarded and had gained national prominence by constructing Rochester's first cyclotron and embarking on an ambitious program of nuclear research. "Look," Lawrence told him over the phone, "we have an important job having to do with the national defense coming up. I can't tell you about it, but I assure you it's very important."

In the days that followed, Lawrence would have that same conversation with every physicist of consequence in the country. These were accomplished scientists—many of them his former students—who were on the brink of exciting careers in nuclear physics. In each case, they dropped what they were doing and came for the simple reason that Lawrence had asked them to. He was so successful in rallying his colleagues to the cause that by November one eminent physicist was joining the staff each day. After he talked to Lawrence, DuBridge was on a train that evening to New York. He felt that "if Lawrence was interested in the program, that was what I wanted to be in."

At a meeting at the Commodore Hotel the following day, Loomis and Lawrence persuaded him to accept the post of director. "DuBridge was such a good sport that he said he would give up everything to do it," said Loomis. He and Lawrence had already started writing down the names of all the talented physicists they could think of. "Then the three of us sat down and made a list," continued Loomis. "The Physical Society was meeting at MIT a few days later, so the recruiting was not done by mail, but in the hallways." It was a cloak-and-dagger affair: some six hundred physicists from all over the country were gathered in Boston for the meeting, which Loomis and Lawrence used as a cover to invite their select candidates to town and ask them to volunteer for the clandestine project. Loomis remembered with wry amusement that the conversations all tended to go more or less the same way: "Well, I don't know if I can or not . . . hmm, well, Jones is coming—oh, is he coming, well, then I'll come. . . ."

For the most part, the conference, which took place during the last week of October, proceeded in a routine manner, and only a very keen observer would have wondered about the hushed conversations in the corridor or the hastily called closed-door sessions on ultrahigh-frequency techniques. Within a very short time, Loomis had his core staff in place. Loomis and Compton hosted a kickoff luncheon at the exclusive Algonquin Club in Boston, which served as the "home away from home" for the British radar specialists and had been equipped with a secret entrance for their use. The two dozen new recruits were brought up to the second-floor dining hall, where they were asked to sign the NDRC's secrecy agreement before being briefed on the new lab and the classified radar work. They would be starting immediately, and within a week many of them would be saying good-bye to their fellow teachers, students, laboratories, and families. Then Loomis formally introduced Bowen to his new American teammates. "It was a grand pep-talk and got the whole group keen to be in on it," DuBridge wrote Lawrence, who had returned to Berkeley and would continue his energetic recruiting efforts from there.

Even Lawrence was caught up in the excitement. On November 1, he dashed off a brief note to Loomis, including a "little sketch" of a design of a power amplifier. "I have been having a rather interesting time with myself thinking about methods of detecting microwaves," he wrote, "and came to the conclusion that the most promising attack on the problem would be to concentrate on the development of radio-frequency power amplifiers along essentially conventional lines. . . ." The next day, he followed up with another letter explaining that he had been talking to Luis Alvarez, who was keen to go ahead with his own experimental design of a "really sensitive detector." Lawrence explained that he was enthusiastic, and even if Alvarez's idea did not ultimately work out, "he will gain valuable experience which would qualify him as an expert in micro wave electronics." Lawrence added, "I am mindful of the fact that I am a novice in the micro wave field and that in all probability to one who has been working in the field for some time my present thoughts about it would be either very elementary or rather beside the point." He was hankering to learn more, however, and had called up Hansen to see if he could talk to him. "I have no doubt that you had a very busy week at Cambridge," he concluded, "and I am anxious to hear the latest developments."

In the end Bowles had to admit that Loomis' recruitment strategy had worked like a charm. Roping Lawrence into the radar project had been a stroke of brilliance. "Loomis was smart as hell," Bowles conceded later. "The Manhattan Project had not yet come into being. Here were all these unemployed nuclear physicists. Why not regiment them? Loomis figured the way to do that was to put Lawrence on the mi-

crowave committee and get Lawrence to pick a head of the lab."
Lawrence tapped DuBridge, and between them they were able to pull in
top physicists like Columbia's Isidor Isaac Rabi, who in turn brought
along two of his brightest students, Jerrold Zacharias and Norman Ramsey. Rabi, who was at the Algonquin luncheon, joined on November 6.
"I was dying to get into something," he later recalled. "When I heard of
this, I said, 'I want to be in on it.' " Ramsey, who was just a few weeks
into a new job at the University of Illinois, joined days later. It was an
extraordinary group. Bowles reckoned Loomis and Lawrence must have
gotten "the whole list—the cream of the crop."

Those who knew Lawrence well in those days, and were familiar with
his ostrichlike ability to bury himself in his work, wondered how Loomis
managed to light a fire under the detached, apolitical physicist. Alvarez,
who had never known Lawrence to show any great interest in world affairs, was struck by the change in his attitude. "[Ernest] was suddenly
converted to the seriousness of the business and saw that physicists
could be useful and therefore should be used—that they could do things
other people couldn't do."

His only contact with the war was when John [his brother] was torpedoed, and I think that's when he woke up and knew it was real; but
again, he had the feeling that it was—John had just been caught in an
alley fight and it was none of his business. . . . I never heard [Ernest]
say anything that indicated that he was seriously concerned with
the war until he got back from talking with Alfred, at which point he
was just fascinated by what he had heard about what could be done
scientifically if you had the motivation and the money and the resources.

While it was true Lawrence never worried aloud about the European
war, or abandoned his work on the cyclotron the way so many of his colleagues quit promising careers to devote themselves to defense work,
Loomis always defended his friend's deep commitment to helping the
British fight the scientific war. "He proved it one hundred percent,"
Loomis would insist years later, "by his reaction—doing everything in
the world to get Luis Alvarez, Ed McMillan, and Lee DuBridge to come
to this new laboratory a year and a month before Pearl Harbor shows
how important he felt it was."

Alfred Lee Loomis, the Wall Street power broker, in the 1930s.

BRAIN WAVES AND DEATH

2

WILLARD RICH

3

Written under a nom de plume, William Richards' thinly veiled novel about Loomis luxurious Tuxedo Park laboratory was published shortly after his suicide in 1940. His sister, Patty Richards, and her husband, Harvard president James B. Conant, worrie the book might cause a scandal. William Richards as a promising young Princeton chemist in the late 1920s.

4

5

Alfred came from a distinguished line of doctors. His
grandfather Alfred Lebbeus Loomis
(*above*) was an eminent tuberculosis
specialist. His father, Henry Patterson
Loomis (*right*), a respected pathologist,
caused a scandal when he walked out on
his family when Alfred was a teenager.

6

7

8

Alfred was raised by his strong-willed
mother, Julia Atterbury Stimson (*above*),
and "fifty aunts, uncles, and cousins," including Henry L. Stimson (*above*), who
was a first cousin. Twenty years Alfred's
senior, he became his mentor.

After his father's early death, Alfred was
determined to take care of his mother and
younger sister, Julia.

9

Loomis married Ellen Farnsworth, "the prettiest girl in Boston," on June 22, 1912, and they joined the fashionable young social set in Tuxedo Park.

10

In 1921, Loomis purchased an elegant gabled mansion in Tuxedo Park, designed by the well-known Philadelphia architect Wilson Eyre; it boasted a cavernous living room and gallery with Jacobean ceilings and elaborately carved mantelpieces where he entertained lavishly.

11

12

During World War I, Loomis' technical genius led to his being made chief of research and development at Aberdeen Proving Ground, where he made major refinements in artillery, including the recoilless "Loomis Shooting Cannon."

13

14

Loomis introduced his sister, Julia, to fellow Yale alumnus Landon Ketchum Thorne (*above*), an up-and-coming bond salesman at Bonbright & Co.; they married in 1911 and had two sons, Landon and Edwin.

Loomis and Thorne became partners in 1919, and the story of their phenomenal success in developing the public utilities business became Wall Street legend.

Julia Loomis Thorne "loved to hold court" at her sprawling Bay Shore estate, Thorneham, and in her jewels and Mainbocher gowns, "looked like one of the royal family."

15

Loomis purchased the enormous Tower House mansion in 1926 and converted it into a luxurious private physics laboratory in the tradition of the great nineteenth century British scientists Charles Darwin and Lord Rayleigh.

16

Loomis' protégé, Garret Hobart, the grandson of McKinley's vice president of the same name, was the laboratory director; his spirited young wife, Manette, was popular with the largely male population of guest scientists.

(Inset) Manette showing off her coquettish side in an amateur theatrical production.

17

18

19

20

George Kistiakowsky, who taught at Princeton with William Richards, was one of the many pioneering young scientists who did experiments at Loomis' Tuxedo lab.

Loomis paid the brilliant and eccentric R. W. Wood, here shown posing in front of the mercury telescope he built in his barn laboratory at East Hampton, to teach him physics.

Loomis and Wood turned Tower House into a "palace of science," and the laboratory guest book shows the names of luminaries from Einstein and Heisenberg to Bohr who made the pilgrimage to Tuxedo Park between 1926 and 1939.

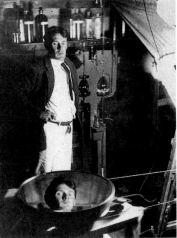

21

22

23

Robert Williams Wood –
Johns Hopkins Un. Baltimore

Karl F. Herzfeld
Johns Hopkins Un. Baltimore

Gano Dunn
20 Washington Square
New York
Sept 26 1926

Malcolm E. Pierce
New York University
Sept. 26, 1926

Richard H. Ranger — Sept 28, 1926.
Radio Corporation of America

Frank Thone
Science Service

Pierre Lecomte du Noüy
The Rockefeller Institute

24

Oct 22, 1926 Frank E. Lutz   American Museum.

Dec. 1, 1926 Garret A. Hobart 3rd.

Jan. 26, 1927.   William T. Richards   Princeton

Feb 7 1827.   L. H. Dawson Naval Res. Lab.

March 27 1927. J. C. Hubbard New York University.
March 27. 1927. E. Schrödinger University of Zurich
June 2 1927 Grace E Ford
June 2 1527 Albert G. Ingalls Assoc. editor. Scientific Am.
July 1927 Bryant Chetter O.F.D.4 New Haven, Conn.

25

December 7, 1935
Myrtle B McGraw N. Y. C.
May 24 1936.
Albert Einstein.
Nathan Rosen
William J Schieber Research Lab.
Sch'dy N. Y. July 30,

In 1931, Loomis and Thorne used their personal fortunes to purchase Honey Horn Plantation, along with twenty-two thousand acres of prime forest land, on Hilton Head Island, South Carolina, as their idyllic private hunting and fishing preserve.

The Ripley's pen-and-ink drawing depicts Loomis and Thorne on horseback, hunting with their hounds.

The brothers-in-law also bankrolled the America's Cup contender, the *Whirlwind* (*center*), shown here racing the *Weetamoe* and *Enterprise* at the start of the trials on June 12, 1930.

The guard tower at Tuxedo Park was designed to protect the exclusive gated colony from outsiders.

29

During a vacation at the Hobart family compound in Maine, Manette was already in love with Alfred, although her husband, later cut out of this photo, "hadn't a clue."

30

31

(*Inset*) Loomis gave each of his sons, Lee, Farney, and Henry, one million dollars to experiment with as teenagers.

Loomis hired the Swiss architect William Lescaze to design the modernist Glass House in 1938, which became a lovers' hideaway for Alfred and Manette. Loomis, his back to the camera, is gazing lovingly at Manette, who is lounging on the patio. In the foreground is Lee, who broke bitterly with his father after learning of the affair.

32

The leading powers in the scientific establishment meet in Berkeley in March 1940 to discuss the giant cyclotron: *(left to right)* Ernest Lawrence, Arthur Compton, Vannevar Bush, James B. Conant, Karl Compton, and Alfred Loomis.

A signed portrait Ernest Lawrence gave to Loomis, who was one of his closest friends and advisers.

The 60-inch cyclotron at the Berkeley Rad Lab: Luis Alvarez is perched on the magnet coil tank next to Edwin McMillan. Standing are *(left to right)* Donald Cooksey, Ernest Lawrence, Robert Thornton, John Backus, and Winfield Salisbury.

One of the many Rad Lab parties at DiBiasi's restaurant hosted by Loomis and Lawrence: Ernest Lawrence, Vannevar Bush, Molly Lawrence, and Alfred Loomis are seated on the left.

Loomis and a group of MIT researchers used a diaper delivery truck, dubbed the "didey wagon," to house one of the first radar speed guns ever built. Loomis had it painted the Tuxedo colors, green and gold, so it would not attract notice.

The historic weekend in October 1940 when members of the secret Tizard mission met with American scientists at Loomis' Tuxedo Park estate. Standing in front of Glass House are *(left to right)* Carroll Wilson, Frank Lewis, Edward Bowles, Taffy Bowen, Lawrence, and Loomis. The British physicist John Cockcroft (not shown) took the picture.

"A pearl beyond price": Taffy Bowen (*left*) shows off the cavity magnetron, the British invention that promised to revolutionize radar, to Lee DuBridge (*center*) and I. I. Rabi.

Rad Lab physicists worked around the clock in the Roof Laboratory at MIT to build the first airborne microwave radar system in 1941.

(*Left to right*) Robert Oppenheimer, Enrico Fermi, and Ernest Lawrence circa 1940. Within two years, they would all be involved in a top-secret project to build the first atomic bomb.

George Kistiakowsky, a member of Loomis' Tuxedo Park circle, would be recruited by Oppenheimer to develop the detonator that would trigger the nuclear explosion.

General George C. Marshall and Secretary of War Henry Stimson both advocated dropping the atomic bomb to bring the war to a quick end with the least possible cost in American lives.

Alfred and Manette were married in Carson City, Nevada, on April 4, 1945, the same day his divorce from his first wife became final. The gossip columns hinted that Loomis had been carrying on with Hobart's young wife for years, and the shocking affair so scandalized New York society that many of his oldest friends turned their backs on him.

44

Manette became great friends with Ernest Lawrence, and a heroic bronze she did of him in 1946 now sits in his Berkeley museum.

45

46

The last of the gentleman scientists: After the war, Alfred Loomis returned to the private life he preferred, though he continued to fund original research and to invite the many Nobel laureates and leading scientists who were his closest friends to join him on all-expense-paid vacations.

WHEN it came to selecting the final site for the new laboratory, Loomis and Bush would team up to outmaneuver the industrialists. At the outset, however, they did not see eye to eye on the subject. Loomis felt strongly that twelve-hour days and intimate cooperation made having one central laboratory the only feasible arrangement, and that it should be based in Washington. This was a radical departure from the military R&D ventures of the first war, however, and already the army was introducing obstacles at every turn. After he received word that Bolling Field would not be available as promised, Loomis tried once again to persuade Bush to base the lab at Carnegie in Washington. Bush refused to budge: "I protested, and we had a hell of an argument that took half the night and a bottle of Scotch."

After mulling over the problem at dinner one night, Loomis and Bowles came to the conclusion that MIT was the only possible site. It was an obvious place to gather academic scientists without attracting attention, and they would be able to get to work much faster because the university would be able to advance the money for the new laboratory until the funds from the NDRC became available. Other war-related projects had been held up for months waiting for the paperwork to be processed and funds allocated. Moreover, MIT already had an independent radar project funded "by an individual"—Loomis himself—who had offered to expedite things by paying the traveling expenses of all the new staff members until such time as the laboratory could do it itself.

Bowles, a die-hard MIT booster, was all for the plan. He wanted all the glory of the radar project to reflect on the school and, by association, on himself. But Loomis knew they would have to tread cautiously. People were bound to raise questions about how MIT had swung such a sweet deal for itself, landing what amounted to the richest research contract of the war. He was a member of the MIT Corporation and a close friend of Compton's. Bush was the ex-dean of engineering. It would seem like blatant favoritism. The real problem, Loomis and Bush agreed, would be Frank Jewett, the head of Bell Telephone Laboratories. He had made no secret of the fact that he thought the lab should go to Bell, which had been asked to run a similar research project for the navy during World War I and could bring far more experience to bear in coordinating such a large and complex industry effort.

As luck would have it, all the principal players happened to be in Washington on October 16, and when Bush learned they were all in town, he called a last minute meeting at his office to settle the question of where to put the laboratory. When the subject arose, Jewett immediately launched into an enthusiastic speech about the virtue of Bell's management expertise as compared to MIT's. "So Jewett made the proper mistake," recalled Bowles, and Bush, "with his usual astuteness, and with Loomis an arch pirate at heart," set the trap.

Bush put the question to Jewett: "What about MIT?"

Sensing he may have gone too far, Jewett tried backpedaling: "Well, it's a wonderful school, it's a wonderful institution . . . but as an educational institution it doesn't know anything about management." Bush and Bowles took umbrage at this, and Jewett, suddenly realizing he had made a blunder by insulting the two MIT men, said, "I didn't mean to criticize the institution, it's a wonderful place. . . ."

At this point Loomis jumped in and cut him off: "Oh, Dr. Jewett, I'm so glad you approve of having the lab at MIT."

Now that this obstacle had been removed, the trick was to present it to the NDRC as a fait accompli. On October 17, the day before the next meeting of the NDRC, Loomis met Compton at the railroad station in Washington and took him back to his suite at the Wardman Park Hotel. There they met with Bush, who told Compton that they had all agreed on MIT and they needed to fix the arrangements at once. After making an emergency call to MIT, Compton reported that he could make available approximately ten thousand square feet of laboratory space and could probably arrange access to a large airplane hangar at the East Boston Airport, which belonged to the National Guard and had a modern machine shop attached. On the strength of this evidence, the contract for the laboratory was voted the following day. One week later, Loomis, Bowles, Wilson, DuBridge, and a handful of others met again at Compton's office at MIT and afterward made an inspection tour of the space set aside for the new laboratory. Loomis then sent an official letter to Bush and the NDRC recommending they take the space at MIT and requesting $138,425 for five development contracts for the equipment they needed. The NDRC approved Loomis' program and allocated the sum of $445,000 to cover the first year of the laboratory's existence.

At the end of October, the pace of events accelerated. Meetings of the microwave committee were now being held at almost weekly intervals, at either Loomis' penthouse in New York, the Carnegie Institute in

Washington, or MIT, where space was rapidly being cleared and the new laboratory would be ready in weeks. Loomis took the wise step of acquainting his new recruits with the leading officers who were in charge of radar development in the army and the navy and hosted a series of luncheons and dinners at the Ritz-Carlton and Wardman Park. On November 1, Frank Lewis, Garret Hobart, and the rest of Loomis' staff packed up all their experimental equipment and took it up to MIT. Two weeks later, after a meeting in Washington, Loomis and Bowen flew back to New York, where a car was waiting at the airport to whisk them to Tuxedo Park. There they met with Tizard, who had returned to Washington for another round of radar talks and had come out to see Loomis one more time before leaving the country. He was deeply gratified to learn of Loomis' progress and, as he wrote, the American firms had "pushed ahead at surprising speed and delivery of a large amount of gear for the first five experimental ten cm airborne sets is expected by 23 November." In a subsequent letter home, he added: "This side of the Atlantic is going to be all-important in a year's time and we shall need to keep in the closest touch."

Also waiting for Loomis at Tuxedo Park was Lawrence, who was winding up his recruiting efforts on the East Coast and would soon be embarking on a cross-country train tour of major American universities, stopping at the University of Chicago and Purdue among others, as he continued searching out sharp minds for the new lab. Roosevelt's re-election in November had helped increase sympathy for Britain, as had the relentless night bombings of London. Most of the scientists Lawrence approached had a burning desire to get involved: many of them were Jews and had close ties to family and friends in Europe. They were shocked by the way the Luftwaffe had laid waste to Coventry and the stories they heard of the Nazi tyranny of Jews, and they wanted to be part of any effort that would bring an end to all that. When the nuclear physicist Ernest Pollard got a telegram requesting his services, he left in such a hurry that five years later, after the war was over, he found the cable still stuffed in the pocket of the lab coat he had abandoned as he dashed out the door. That none of them knew a damn thing about radar did not seem to matter. "They were light on their feet," said Alvarez. "They knew something about electronics because of their work with accelerators, but the real reason was that they were the best people and they were adaptable to anything."

Loomis shuttered his Tuxedo Park laboratory for the winter, though he probably guessed he would be away considerably longer than that. He bade good-bye to his wife, who had been quite unwell again that summer and was planning to move back to her parents' home in Dedham, Massachusetts. She would be well taken care of there, and as it was not far from Boston, Loomis would be able to look in on her now and then. All three of his young sons had enlisted: Lee and Henry were in the navy, and Farney, who had graduated from Harvard Medical School, had joined the army's 9th Mountain Division. According to Loomis' grim calculations, which he ill advisedly shared with a family member, "the odds were one of them would not return." Only Manette, with two little boys to look after, would remain on in Tuxedo Park. Loomis would arrange to make frequent visits to New York to be with her, always meeting her at the Glass House, and would continue the affair in secret throughout the war years.

After making arrangements for the last of his valuable equipment to be shipped to MIT, Loomis, with Bowen and Lawrence in tow, hurried up to Cambridge. For the next four years, he would drive himself and his band of physicists almost without break to develop the all-important radar warning systems based on the magnetron. Looking back on that tense autumn in 1940, Bowen boasted, "It was a gift from the gods we disclosed to Alfred Loomis and Karl Compton."

Few understood better than Bush the critical role this unprecedented partnership would play in determining the course of the war. The British and American physicists had joined together to beat the Germans, and their collaboration immediately resulted in a more effective war effort and contributed significantly to both nations' ability to gain an edge on German science. The cooperation among scientists would later extend to military men and would have striking results in the development of new radar devices and their performance in the field of battle. It was not always easy. The two sides had quarrels that were, like the disagreements between Churchill and Roosevelt, "the quarrels of brothers." But as Bush later observed of the Tizard Mission: "Much has been written about the disagreements between allies during a great war. Little has been written about the deep friendships which appear between comrades in arms of different nations, even among comrades whose efforts, behind the lines, are devoted to placing advanced weapons in the hands of fighting men."

Chapter 10

THE BLITZ

He was also trying to run things his own way: was there
anything queer about that? Maybe. Maybe, though, it was only
the behavior of a man who was used to giving orders.

—WR, from *Brain Waves and Death*

O N November 11, 1940, Armistice Day, Loomis held the first meeting of the radar lab in its new headquarters on the ground floor of Building 4, a squat concrete structure on the edge of MIT's campus. About twenty people gathered into one of the small classrooms, all of them having received an invitation from Loomis that was so worded "as to sound like a courteous order." Security was tight. The windows of the laboratory were painted black, and a guard was posted at the door. Only a few of the physicists who had been recruited thus far managed to make it for opening day, but everybody found they knew everybody else. As Rabi later remarked, "We all came from the same bar." It felt like a family reunion of sorts, and people were already walking around squabbling good-naturedly about where their benches should go and what they needed to buy. It made for an easy informality, as well as a sense of high spirits and fun.

Karl Compton opened the meeting by providing the early arrivals with a general overview of the situation and then turned the meeting over to Loomis, whom he introduced as "the man who knows more about radiolocation than anyone else in America." Loomis filled them in on the fundamentals of microwave radar and outlined the first problem to be tackled by the group: an "airborne interception," or AI, radar to defeat the night bomber. Everyone present knew that Göring had changed his strategy and that since the beginning of November the Luftwaffe's daily assaults on England had been replaced by night attacks.

The British had long anticipated the change in strategy, and it had been the primary concern of the Tizard Mission. The first German air raids, which had begun on August 8 and had rapidly increased in intensity, had been directed at RAF bases. Since September 7, mass raids had been ravaging London and other major cities in England. Thanks to the Chain Home system, the British had been able to spot the incoming planes and had exacted a toll. During August, the Luftwaffe's losses in raids over England was 15 percent—in all that month, they lost 957 aircraft. In the great air battles of September, the Germans had lost 185 aircraft out of an attack force of 500.

According to Bill Tuller, one of the young MIT researchers who had been at Tuxedo Park, Loomis, in his usual straightforward manner, outlined the challenge before them: "At that time, day bombing had just become too costly, and night bombing was beginning to be used by the Germans with striking results. The problem then was the detection of the night bomber by an operator in the night fighter, who was then to guide the fighter pilot to a position from which the bomber could be seen by the dark-adapted pilot."

These night attacks were forcing the British pilots to rely increasingly on the still crude airborne interception radar that had been developed by Bowen and his compatriots in the preceding two years. Because of the limitations of AI, the British had developed a whole different technique called "ground-controlled interception" (GCI). In this system, a controller on the ground, watching the air situation on a special radar set, used the low-frequency Chain Home stations to pick up invading aircraft. The high-frequency GCI radars, which were more accurate but shorter range, would then target a German bomber and give detailed vectors to the fighter plane under his control, maneuvering the plane into position one to three miles behind and just below the target. The radar operator in the plane was then instructed to "flash his weapon," and the airborne radar system took over, guiding the fighter to within visual range.

During the daytime raids, all that was required was to bring the pilots into the general vicinity of the incoming stream of bombers, and then the pilot took over, using his own sight and judgment to select targets and gun down the enemy. The nighttime raids, however, demanded much more accurate course directions than these radar sets could deliver. It also relegated experienced pilots to the role of hapless chauffeur right up to the moment when they were close enough to see the blur of the enemy plane against the sky; only then, at the last minute, could they press home their attack.

It went without saying that this complicated system required a very high order of skill and virtuosity on the part of the ground controller, the radio operator on the plane, and the pilot, and that in wartime Britain there was a critical shortage of such talent. In addition to the radar system's technical shortcomings, the RAF's Blenheim bomber planes lacked the speed and weaponry required to take on the Luftwaffe and were having little success after sundown. The radar system that had enabled the RAF to function so brilliantly during the daylight raids of the Battle of Britain left them blundering in the dark. The night fighters desperately needed more sharply defined radar beams, and that meant microwaves. The magnetron promised the development of radar sets using much shorter wavelengths than the 1.5 meter then in service. Only at wavelengths below 10 centimeters could an antenna be small enough to be installed on an airplane yet still produce a sharp enough beam to give a highly accurate read on the enemy's location. This electronic eye—which could see through clouds, fog, and cover of night—was Britain's most pressing radar need, Loomis told them, and the laboratory's first priority.

After Loomis' briefing, the various industry representatives on the microwave committee gave updates on their progress in manufacturing the components. Although the original timetable had been extravagantly optimistic, there were no delays. Incredibly, almost all of them would meet the deadline. The following day, Loomis and Lawrence declared the new lab open for business. Bell delivered the first five copies of the British magnetron right on schedule. After everyone had a chance to admire them, they were locked in a safe in DuBridge's office.

Frank Lewis, one of the young MIT researchers who had joined

Loomis' Tuxedo operation, had been the first to arrive and along with several other members of the original staff spent the first few days giving demonstrations of their microwave aircraft detector, mounted on the Loomis Laboratories truck, to the new recruits at the East Boston Airport. "Loomis brought the entire crew from his lab at Tuxedo Park," he recalled. "We picked up all the equipment we had originated, and all of the stuff we had bought to work with, and we put it into the didey wagon and we drove the truck up there." Having already heard reports of the British tracking systems, and the two-hundred-mile range covered by their equipment, some of the new recruits scoffed at the Tuxedo device's measly two-mile range. Young, cocky, and supremely confident, they were certain that with their collective brainpower they would invent a radar system that would whip the Nazis and win the war.

The first week in Cambridge was tense and bewildering. Loomis was creating the radar laboratory out of thin air, and the fact that it did not really exist yet, combined with all the secrecy surrounding the project, made it hard to know exactly how to proceed. Everyone was told the project they were going to work on was for the military: "We had to keep our mouths zipped shut all the time," recalled Lewis, "and we had to be sure that we were working with people who were cleared for this." As they began to set up shop and contact the various manufacturers about the delivery of parts and supplies, they realized for the first time the extent to which Loomis had simply willed the enterprise into being. Seeing how far the country still was from entering the war, and knowing from firsthand experience how difficult it was to move the services in a new direction, Loomis had essentially hijacked the project for himself: "The microwave committee, which was a fictitious organization that was set up by Alfred Loomis, had made arrangement with all these government contractors to work on these problems," explained Lewis. "They had no official appointment from the federal government to do this. But Loomis got them all talked into doing it, and they were so convinced that they were it, they went right ahead. And it's a good thing they did."

For all Loomis' wealth and freewheeling style—he had used money out of his own pocket to jump-start the new lab—the government had him on a short leash. The feasibility of microwave radar had to be established quickly if the project was to get the green light and receive further funding. "Loomis got all the radio manufacturers that he could get his hands on to buy the idea that this was going to be a big show, and he would be the main propellant, and they'd better do what he told them," said Lewis. "He didn't put it in those words, but that's what he was saying." As far as the salesmanship was concerned, "It took the talents of a Loomis and a Compton," agreed Bowles, adding that there was "more than a bit of skullduggery" that went into the early contracts, most of which were negotiated verbally and not set down on paper until months later, creating all sorts of havoc. "We pretty well got away with murder."

Everyone agreed that the new lab needed to have some sort of title, but a descriptive yet nonrevealing name was hard to find. Finally, one of the Berkeley group suggested calling it the MIT Radiation Laboratory in honor of Lawrence, who was largely responsible for their all being there. The misleading name would account for the large and rather sudden concentration of experimental physicists and cyclotroneers in Cambridge, while at the same time it would be descriptive, in a sly way, of their purpose. In the interests of secrecy, they also hoped the disguise would fool outsiders into thinking that they were engaged in research as altogether remote from the war effort as nuclear fission, which was considered of no practical significance as compared to radar. The "Rad Lab" met with unanimous approval and was officially adopted.

As the parts began to appear, and more physicists arrived, a loose structure took form. The immediate work was divvied up into seven technical sections based on the components, and as everyone had expertise in different fields but not specifically in the radar set's dissembled parts, the selection process was somewhat random. "We chose up just like a baseball team," said Rabi. "We chose up sides. What would we take?" Turner took receivers; Bainbridge took pulse modulators; Lewis took klystrons; and so on. Rabi opted for the magnetron, though as he recalled later, "I had no idea how it worked." He was hardly alone. Most of the young nuclear physicists, freshly arrived from their university laboratories, knew little to nothing about microwave electronics and had only the vaguest understanding of how the British ten-centimeter magnetron would transmit power for a radar set. But then no one else did, either. Microwave radar was virgin territory and the magnetron brand-new technology.

Rabi was confident that he and his fellow physicists had "the intellectual mobility" to find out everything they needed to know. As he was in charge of the magnetron group, he decided to go around to MIT and ask some of their electrical engineers for advice. "After talking to them I could see they didn't know anything, either," he said, "so we started absolutely fresh and designed magnetrons." Everybody did the best they could, hopping back and forth across organizational lines as needed and throwing out ideas to members of one group or another. Caught up in the excitement of the adventure and imbued with a sense of their own importance and assured success, they plunged into the unknown.

"It's simple," Rabi boasted in one of the early sessions, when they were seated around a table, staring at the disassembled parts of a magnetron. "It's just a kind of whistle."

"Okay, Rabi," challenged Edward Condon, one of the Berkeley physicists recruited by Lawrence, "how does a whistle work?" The long silence before Rabi attempted an answer spoke volumes about how much they still had to learn.

Bowen was struck by the easy camaraderie that prevailed at the lab: "Everyone worked long hours and did not spare themselves. Here was the cream of American scientists, hell-bent on doing all they could for the war effort." There was little time for relaxation, but on Friday nights the gang all gathered at the bar behind the Commander Hotel just back of Harvard Square, where Luis Alvarez and Ed McMillan, among others, were staying. Inevitably, this was soon dubbed "Project 4," a last and vital addendum to the lab's must-do list. Bowen generally took a lot of ribbing on these outings, particularly because the mural decorating the walls of the bar depicted various patriotic scenes "dear to the hearts of all Americans"—the Boston Tea Party, Paul Revere's ride, and Minutemen shooting through the trees at the Redcoats. "The message was loud and clear—this was where they beat the pants off the British," he recalled with amusement. "Proceedings usually began with an expression of solidarity, a friendly toast to the British—'The hell with the Limeys.' " There were also weekly dinners in Chinatown, limerick competitions, and the "laugh meter," featuring jokes only physicists could appreciate. As a break from the tension, everyone read science-fiction novels and dog-eared copies were passed around.

Loomis moved into the Ritz-Carlton Hotel in Boston, where he occupied a lavish suite—particularly given the spartan dormitory quarters assigned to his physicists—and often hosted private dinners for Bush and other NDRC officials. Because of his long hours at the lab and erratic travel schedule, Ellen spent most of her time with her parents in

Dedham, away from the noise and dirt of the city. Because Hobart was also immersed in the radar project, and both her boys were now enrolled at the Fay School in Massachusetts, Manette had an excuse to be in Boston and came as often as she could. She and Loomis continued to see one another secretly and were so circumspect that it seems neither their families nor anyone at the lab guessed what was going on. They were greatly helped by the fact that Boston had been flooded with hundreds of newcomers, and people's lives no longer conformed to a regular pattern. Young men were scattered in hotels and temporary housing all over the city, many of them separated from the wives and children they had left behind. Young women who had never worked before were taking jobs in town and going about in slacks and sweaters. There was a general feeling of chaos and things building toward a crisis, and it tended to make people more casual than they might have been in more orderly times. Later, the dim-outs made it hard to get around at night, and the darkness no doubt covered a multitude of sins.

THROUGHOUT that fall, work proceeded at a furious pace that DuBridge, with heavy irony, described as "the blitz." The physicists set about trying to understand why the magnetron worked so well and initiated theoretical and experimental studies. Rabi's group quickly discovered that the magnetron could produce far more power than the British had suspected and soon enough was known to improve the efficiency considerably. As the parts were delivered, work was begun on testing and adapting them, while other groups designed components for use in an aircraft.

By mid-December, the lab had almost doubled in size and was employing thirty-six people: thirty physicists, three guards, two stockboys, and an indomitable secretary by the name of Edythe Baker, whose idea it had been to paint the windows. It had also outgrown its original space, first moving upstairs, via a spiral stairway, to the second floor and then upward again to the top of Building 6. There they had erected a crude wooden "penthouse," about twenty by fifty feet, covered in green tarpaper, and a second story was already being added. The so-called Roof Laboratory soon became the main hub of activity as the various components they had ordered started to arrive and they began to hitch up the radar system.

On December 16, emboldened by their progress, Loomis wrote out on the blackboard a schedule of ambitious target dates for the AI project:

Goal 1: By January 6, a microwave system working in the Roof Lab.
Goal 2: By February 1, a working system mounted in a B-18 bomber supplied by the Army.
Goal 3: By March 1, a working system adapted for an A-20A attack bomber (the plane most likely to be used for night combat).

Loomis and Lawrence set up a group dedicated to assembling the system and appointed Alvarez expediter to ensure that the deadlines would be met.

By late December, a complete ten-centimeter pulsed microwave radar set was assembled on the roof, and they would soon be able to begin testing it against buildings in Boston, just across the Charles River. This was a two-antenna system, with separate parabolas for transmitting and receiving. The two large dishes, mounted in a rickety apparatus in the Roof Lab, looked like two monstrous black eyes staring out at the golden dome of the State House and the Boston skyline. The physicists, shivering in their unheated penthouse laboratory, had put together the parts and wrestled the unwieldy system into operation. Now the only question was, Could their electronic eyes see?

Loomis took a quick break for Christmas, heading to Hilton Head for the holidays. All three of his sons were away at war, but Ellen would be there, along with Julia and Landon Thorne and their family. He invited along the Comptons, as well as Bowen, who was on his own for the holidays, as his wife was stuck back in England owing to wartime travel restrictions. Loomis knew the Welshman had more than earned a rest after an exhausting few weeks demonstrating the British AI and ASV (air-to-surface vessel) Mark II long-wave radar sets to American military personnel. Bowen had showed off the performance of the Mark II, which he had helped develop and which had been fitted into a U.S. Navy PBY aircraft. Flying over merchant ships in the Atlantic, he had successfully picked up an echo from a capital ship at a range of about sixty miles. Satisfied that this was the equipment to adopt, the navy had finally agreed to take over for the Tizard Mission, placing orders with the Philco Corporation for seven thousand copies of the ASV radar systems. An additional ten thousand sets were ordered from Canada's Research Enterprises Ltd. Ultimately, the procurement would run into hundreds of thousands of sets, the majority of which would be used by American forces.

Just before New Year's Eve, Roosevelt gave a speech promising aid to Britain from the "Arsenal of Democracy." Bowen, who was heartened by the president's address, only hoped it had not come too late. As Loomis and his friends rang in the new year, they had much to celebrate and much yet still to do. Loomis' best estimate was that the project would take at least two years and millions of dollars in government funds. England was being methodically bombed by the Luftwaffe. The dark winter of the Blitz had begun.

O N a cold, clear morning on Saturday, January 4, 1941, two days ahead of schedule, a radar beam was sent out from the Roof Lab, and the first echoes from the Christian Science church tower in Boston were detected. In less than eight weeks after they walked into the Rad Lab, Loomis and his band of microwave novices had managed to build a working prototype of a radar system. It was far from perfect; in fact, there were so many tuning stubs introduced at so many different points that coaxing the tuning was "distinctly an adventure." But it was a start. An excited DuBridge telegraphed Lawrence at home in Berkeley, where he had returned just in time for the birth of his second son on January 2:

> ROOF OUTFIT IN FALL [sic] SWING LOOMIS IS
> JUBILANT. . . . FEBRUARY FIRST DATE LOOKS EASY
> IF SHIP COMES IN HOPE YOU ARE AS PROUD OF
> YOUR BABY AS WE ARE OF OURS=LEE

Loomis' euphoria was short-lived. It quickly became apparent that there were almost as many things wrong with the system as there were right. To begin with, it was poorly designed and was altogether too large and unwieldy. The main problem still facing them, and one that seemed to have no easy solution, was that the radar system they were charged with designing had to be compact enough to fit into the nose of a fighter plane. For the system to be small enough to be practical, they would have to find a way to use a single antenna, or "duplexer," for both transmitting and receiving. Unfortunately, no one knew how to build such a device.

An air of gloom descended on the Rad Lab. In a sense, they were back to square one. The design of the duplexer—what the British called a "TR" (transmit-receive) box—had confounded them from the beginning. Without a duplexing or switching device, or some kind of protection, the main transmitted pulse would burn out the receiver crystal. Because the outgoing signal was a million times stronger than the incoming echo, the question was how to use a single dish that poured out a powerful radar beam without swamping the feeble echo that bounced off the target and returned in a few microseconds.

DuBridge organized several teams to attack the problem. From day one, the physicists at the Rad Lab were guided by Loomis' insistence on a hands-on approach and practical rather than theoretical solutions— they lived by the law of "cut and try." Finally, after some frantic efforts and several failed attempts, a makeshift single-antenna system was made to operate. The team, under Jim Lawson, who happened to be an amateur radio enthusiast, succeeded in fashioning a TR box by using a klystron amplifier as a buffer in the line from the antenna to the receiver crystal. As Alvarez observed, "If we had been paid in proportion to our contributions to the success of the first microwave radar program, Jim Lawson would have earned more than half the monthly payroll."

Lawson's roof team spent hours on end fiddling with the homemade contraption, one of them working in a bulky coonskin coat against the cold. The signal-to-noise ratio in the receiver made it very hard to stay on the proper frequency. One day, while they were working with the system, they picked up a great deal of interference that made it impossible to pick out a signal. "We nuclear physicists had absolutely no idea what to do," recalled Alvarez. Then one of the Berkeley group asked if anyone had a pair of earphones. An MIT engineer ransacked the classrooms and finally found an old pair in a student laboratory. When they hooked up the earphones to the radio receiver, they heard a voice crackling: "Hello CQ, CQ, hello CQ." The mysterious noise they had been hearing was a local amateur radio operator announcing himself over the airwaves. After taking some added precautions, Alvarez noted, "We were back in the radar business again."

On a raw New England morning on January 10, the single-dish system finally picked up echoes from buildings in Boston. DuBridge, who was in Washington, got the news in the form of a cryptic telegram— "HAVE SUCCEEDED WITH ONE EYE"—just in time to inform a meeting of the microwave committee that was being held that day.

They were not out of the woods yet. The makeshift system was subject to frequent breakdowns, and while they had managed to obtain a signal by pointing the dish steadily at the target, a practice called searchlighting, the weak signal produced by their equipment did not seem to hold out much promise for scanning. For that, they would need still more power. As repeated attempts to pick up airborne signals failed, and frustration mounted, many of the Rad Lab physicists began to doubt that the rooftop system would ever perform this essential feat. As Bowen later recalled, "For the first, and possibly only, time a mood of pessimism crept into the group and some doubts were expressed about whether a system would ever be capable of receiving echoes from an aircraft."

The deadline pressure was exacerbated by the tension created by the war and the political demands from Washington. Things could not move fast enough for Bush. "It was characteristic of Bush's management," Bowles observed later. "He wanted results. He was constantly putting the blowtorch to us." Loomis was under enormous pressure to succeed. A great deal of jealousy had been aroused on the part of industrialists who saw the Rad Lab as future competition and were critical of the group's fitness for the job. There was also considerable skepticism on the part of some government officials who questioned whether a bunch of academic physicists—whom the army and navy derisively called "doubledomes"—could successfully carry out such an urgent wartime project. If it turned out to be a wild goose chase, who was going to answer for the wasted taxpayers' dollars? Even respected members of the scientific community expressed reservations. Earlier in 1940, the grand old man of American physics, Robert Millikan, had warned that it was "a mistake . . . to concentrate fifty prima donnas in physics at any one spot."

The February 1 deadline to have a working system mounted in a B-18 bomber came and went. Even Lawrence, visiting the Roof Lab four days after the target date, wondered aloud whether the magnetron could ever be made into an operational airborne unit. Alvarez, who was taking the maestro on his rounds, defended the rooftop team's efforts and bet Lawrence they could probably make it work well enough to detect a signal from a flying aircraft. They shook on it, and Lawrence told him he was prepared to eat his words. They had two days before he and Loomis were due back in Washington to give a progress report to the microwave committee.

Alvarez and his team worked around the clock all the next day but came up empty-handed. On Friday, February 7, the microwave committee gathered in one of Bush's conference rooms at Carnegie. The more impatient members expressed dismay at the lab's lack of progress and suggested junking the whole project. Loomis and Lawrence had pressed their luck as far as they could, and the army and navy were eager to see the last of the scientists and their microwave radar.

Early that same morning, there was a flurry of activity at the rooftop lab. Alvarez, a Berkeley colleague named Lauriston Marshall, and a handful of others were making a last ditch effort to track a plane and prove the project's viability in time for the meeting. One of the physicists, operating on a hunch, decided to detach the parabola and hold the antenna by hand. Another scouted for planes, peering through a crude telescope. As he had done for days now, Alvarez gazed dully at the scope. Suddenly "a blip appeared." Marshall stared at the monitor. As the plane gained in altitude, it was tracked on the screen. Twisting his head to look out from the rooftop lab, Alvarez saw a commercial plane disappearing into the distance. He scrambled down the penthouse's narrow spiral stairway and raced to the phone.

When the call was announced, the meeting fell silent. DuBridge took the receiver and heard Alvarez and Marshall excitedly blurt out, "We've detected an airplane at two miles." DuBridge held up two fingers. Lawrence "caught it right away," his grin showing that he understood that they had obtained echoes at a range of two miles. For the benefit of the rest of the committee, DuBridge reported: "We've done it, boys."

Lawrence telegraphed Alvarez: "I HAD MY WORDS FOR LUNCH."

The microwave committee voted confidence. With the rooftop AI system up and working, Alvarez and McMillan—Lawrence's two hand-picked protégés from Berkeley—immediately set to work designing and building a wooden mockup of the bombardier's compartment in a B-18. Alvarez installed a second AI system in their wooden prototype, which was equipped with a special Plexiglas nose that was transparent to microwaves. Between February 13 and March 5, they worked over their test until the Douglas B-18 that had been assigned to the project by the U.S. Army Air Corps finally arrived. After extensive ground tests—actually roof tests, using a water tower on a building six miles away— McMillan declared that the experimental airborne ten-centimeter radar system was ready to fly. The B-18, which had been flown up from Wright Field by an army crew, was waiting at the National Guard hangar at Logan Airport. On its first outing on March 10, the radar system performed with mixed results, but after several weeks and many modifications, its performance gradually improved.

On March 27, Alvarez and McMillan headed out for the first test run using aircraft as a target, taking along Bowen as an observer. They proceeded eastward over Cape Cod in search of open skies and were surprised at how clearly the ships below showed up on their radar screen. This was a far better result than had been expected. They then made several runs at the target, a single-engine plane borrowed from the National Guard, and got satisfactory echoes at a range of two to three miles. They were feeling quite pleased and were about to turn back when several large merchant ships in Cape Cod Bay caught their attention. Switching off the elevation scan and leaving the radar set to give range and azimuth signals only, they did a run at a large ship in the bay. Flying at about two thousand feet over the water, they tracked a ten-thousand-ton vessel heading for port. While there is no record of the maximum range at which it was detected, Bowen estimated it was about ten miles. More important, the system's admirable performance had not been hampered by the "sea return," the interfering echoes from the ocean's surface. While he knew this was not exceptional compared to the British long-wave ASV radar, "for the first flight of a centimeter-wave radar it was a great performance." Over the roar of the bomber's engines, Bowen could hear his colleagues' wild cheering.

Unable to resist the temptation to try for the extra mile, Bowen said, "Let's go to New London and see if we can find a submarine." New London was home to a major navy submarine yard, and they could be there in thirty minutes' flying time. Just a few days earlier at Tuxedo Park, Bowen had been talking to Loomis about the pressing need for a microwave sub-hunting radar, so this seemed as good a time as any to test the system's potential. Barely able to contain their excitement, Alvarez and McMillan agreed at once and instructed the pilot to head down to Connecticut. As they flew low over Long Island Sound, the radar picked up several large submarines cruising offshore. One was fully surfaced and presented an excellent target. They made several runs broadside on and obtained a strong signal at a maximum range of four to five miles. For the scientists aboard the plane, it was a dramatic moment—no one had ever detected a submarine with airborne microwave radar.

Their sightings were the first real evidence that radar performed well over water. It was "an epoch-making flight," Bowen wrote in his memoirs. "We returned in triumph and the news spread around the Laboratory like wildfire."

From then on, ASV radar for submarine and ship detection was added to the Rad Lab's growing roster of projects and would soon become far more important than their original assignment, as by this time the Battle of Britain was ebbing and the British had lost interest in the night fighter. As Bowen, Alvarez, and McMillan had observed, their airborne microwave radar was perfectly suited for submarine detection, which was a lucky break for the British. "After the Luftwaffe retired from the Battle of Britain, German bombers had only a nuisance value," recalled Alvarez. "The German submarine campaign against Allied shipping, on the other hand, could starve the British to the point of surrender." It would take only minor alterations to turn the airborne system into a highly successful ship detection system, but the tactical advantages were immense. It was a whole new kind of radar and an entirely different breed of defensive weapon. At that very moment, German submarines were beginning to appear in U.S. waters near the East Coast and were harassing the vital transatlantic freight route. The list of sinkings on the Atlantic highway was horrific—over four million tons by the end of 1940—and it was becoming very clear that England would not be able to hold out much longer unless some defense was found. If the Germans were to continue successfully to disrupt Allied shipping, they could defeat the British Isles. While America was not yet in the war, the U.S. Navy realized that airborne microwave radar provided them with a means of detecting this dangerous threat. The navy immediately ordered a trial system, and the British wanted two sets as soon as they could get their hands on them. Here were the first fruits of the Tizard Mission.

For Loomis and the Rad Lab physicists, March was a turning point. For four months, all of their efforts had been focused on building a working radar system and getting it aboard a plane within the time frame Loomis had mapped out on the blackboard in mid-December. They had accomplished that and much more, all of which was described in detail in Loomis' first report on the lab, which the microwave committee had submitted to Bush at the beginning of the month.

In short order, the NDRC approved another $300,000 for the lab, and it was estimated that more than $1 million would be needed for salaries to prolong the work another year. When Congress was slow to approve the funds, threatening to stall the lab's progress, Loomis and Compton pulled one of their end runs, first convincing the MIT Corporation to come up with $500,000 and then appealing to their old friend John D. Rockefeller Jr., who agreed to help underwrite the salaries of the technical staff to the tune of $500,000. Private enterprise, in Loomis' view, could move mountains in the time it took the government to pass a single bill. With the threat of war looming, however, Congress was eventually persuaded to fork over the money, and both MIT and Rockefeller were repaid.

As Loomis continued to conspire behind the scenes to keep the lab afloat, and to agitate for preparedness among the power elite, he chafed at Roosevelt's reluctance to publicly back Britain's cause. Ernest Lawrence, in a letter to Robert Sproul, the president of Berkeley, recalled Loomis' insistence that research for war required speed, and Congress' hesitancy, coupled with the military's intransigence, could cost them dearly:

> He drew a striking parallel between the present international situation and the financial situation prior to the crash. He said that now people are asking him when we will enter the war just as in 1928 his friends were asking him when the stock market crash was coming. He said that in both cases such a question is quite beside the point. He said that once a person admitted a stock market crash was coming a prudent individual will immediately get out of the stock market and not consider when the crash is coming and thereby try to hang on and make some more profits. Likewise at the present time it is of secondary importance when we will get in; of first importance is the admission that we are going to get in, and our action accordingly should be that of preparing just as though we were actually in the war!

With that in mind, Loomis stopped at Woodley on April 21 for a long overdue visit with Stimson, whom he had not seen for some weeks. As usual, Loomis used the opportunity to lobby for the importance of the new radar detectors, which the Army Signal Corps was still fussing over and finding every excuse not to embrace. Throughout that winter and spring, Loomis' anxiety over America's slow pace in preparing for war had increased—it was not nearly as much as he and his colleagues had urged. After the desperate air battles fought in the British skies the previous summer, the defeat of the Luftwaffe had been followed by a strangely quiet winter in the European war. While there was little doubt that Hitler would mount another campaign that spring or summer in a final effort to conquer the British Isles, and the German U-boats were already advancing his cause, it was still difficult for most Americans to face the fact that the country might have to intervene. To prop up Britain, which was faltering, and to keep the country out of the war, the administration had enacted the lend-lease bill, allowing Britain to borrow war supplies against the promise to repay after victory. The agreement was, Churchill wrote Roosevelt, "a statement of the minimum action necessary to achieve our common purpose." But it touched off a long, bitter debate in Congress and was eventually passed in March.

Despite the controversy, Loomis shared the secretary of war's impatience with the isolationists and the president's overly cautious course, which appeared to be one of waiting for circumstance to start the fight for him. Stimson argued that if the policy of sustaining Great Britain was to succeed, America had to throw the major part of her naval strength into the Atlantic battle. There was simply no other way to ensure the safe delivery of the lend-lease supplies. Both Loomis and Stimson respected Roosevelt's political acumen, but as Stimson noted in his diary, they believed the president should take more decisive action, and if he said frankly that force was needed, and asked for the country's approval, he would be supported:

I found both [Harvey] Bundy [Stimson's liaison to the War Department] and Loomis at the Department and I spent a large part of the morning talking to them. . . . I found everybody rather discouraged by the war news and by the fact that the President doesn't seem to be keeping his leadership in regard to the matter. There has been one of Walter Lippmann's articles in last Saturday's papers which gives the situation as a great many people are thinking it. It's rather a defect in his tone and attitude when he does discuss the matter in his press conferences that is the cause of the trouble. We are in such a serious situation that I think people feel that it is no time to joke about it and yet the President's press conferences are always on a light tone. I found that complaint quite universal—that he had not taken a serious enough note with the people. . . . Alfred Loomis was at lunch and dinner with me and it was very good to see him again and to talk with him. He gave me some very encouraging news about the progress of his work in Boston for the defense matters and he told me that the great victory of the British in the Mediterranean Sea a short time ago was due to their being able to locate with a new device the Italian ships in the dark.

During Loomis' visits, Stimson often sought out his advice on various advanced weapons being developed by the services, and on this occasion he was eager to talk to him about a new device that recently had come to his attention. Drawing on Loomis' background in the Army Ordnance Department during the previous war, Stimson wanted to know "if there was a way of using our new bantam cars with a good-sized gun in them to stop German tanks." McCloy had suggested putting a tank-killing gun in one of the new little jeeps, and General Marshall had informed him they were working on a similar idea in connection with airplanes—the cars were light enough to be transported by a big aircraft "two at a time." Such was Stimson's faith in Loomis, and lack of confidence in the originality of his forces, that he asked his cousin to "apply his inventive head" to the problem and to accompany him to Fort Knox to see them in action. "These little cars will run everywhere and run very fast and are typically American because they have the flexibility which appeals to the initiative of the young."

On May 6, Stimson delivered a radio address supporting active naval assistance to the British, stating as clearly as he dared his conviction that war was coming: "I am not one of those who think that the priceless freedom of our country can be saved without sacrifice. It can not. That has not been the way by which during millions of years humanity has slowly and painfully toiled upwards towards a better and more human civilization. The men who suffered at Valley Forge and won Yorktown gave more than money to the cause of freedom. Today a small group of evil leaders have taught the young men of Germany that the freedom of other men and nations must be destroyed. Today those young men are ready to die for that perverted conviction. Unless we on our side are ready to sacrifice and, if need be, die for the conviction that the freedom of America must be saved, it will not be saved. Only by a readiness for the same sacrifice can that freedom be preserved."

While Stimson was not the only political leader to express this view, it was one of the boldest speeches by a cabinet member at the time. Roosevelt, however, continued to listen to the contrasting advice of his State Department advisers, and Stimson's diary entries over the next few weeks reflect his growing pessimism "that the country has it in itself to meet such an emergency." Loomis, who had completed his assignment to study tanks, returned two weeks later, bringing with him a report and some photographs of a trial of the Bantam cars conducted by the cavalry of the 1st Division at El Paso. Stimson was delighted to learn that the idea he and McCloy had come up with had been proving successful: "The tests showed that the gun thus mounted was the realization of what everybody is trying for now—a moveable gun mount. The car is very speedy; easily maneuvered; and the gun has been put on it by these Cavalrymen without any difficulty and with it they made much better scores."

On May 11, London suffered its worst air raid of the war to date, with more than 1,400 killed. Three days later, the great British warship *Hood* was sunk. On May 22, Bush called on Stimson, and they had a long conference concerning Bush's desire for a new organization for scientific research for the army and navy:

> He told me that the Navy needed it much more than the Army— they were more backward in it—but that the Army needed it somewhat. He told me that his proposition was that a new Assistant Secretary of War and Secretary of the Navy should be created which had this in charge, and when I asked who he recommended for it in regard to the Army, he said, "Alfred Loomis."

Five days later, on May 27, the president gave a vigorous radio speech that, while falling well short of what Stimson had suggested, firmly asserted the doctrine of the freedom of the seas and made it clear America intended to use "all additional measures necessary" to assure the delivery of supplies to Great Britain. Roosevelt also declared an "unlimited national emergency," giving his administration broader powers in dealing with the crisis.

On June 5, Loomis and Stimson had lunch at Woodley to discuss the situation and Bush's proposal in particular. "We hammered out the various ways and methods which he would have to do in his work," wrote Stimson. "On the whole I think it is a very satisfactory arrangement, or will be one." Loomis continued to frequent Woodley in June, and predominant in all of his talks with Stimson was his message, which he stressed over and over again, that many of the Rad Lab's new airplane detectors were ready and should be put into use as soon as possible. A few days later, Stimson noted in his diary that after dinner with Loomis, he and Bundy "talked over Alfred's particular specialty and what we should do to get the better system of communications and the protection system into the Army." Bundy, who was a lawyer with no background in science, did his best to maintain good relations with the military while trying to help clear a path for the scientists. Whenever they hit a roadblock, he later recalled: "Bush would needle me and then I would needle the secretary and then the secretary would hit the army over the head."

The only problem was that the army knew that Stimson's sympathies lay with the scientists—with Bush, Compton, and Loomis—and that he, too, was impatient with them for failing to modify their weaponry and methods as soon as the new technical advances became available. Both sides had nothing but harsh opinions of the other. As Bundy put it, "The military don't like to be needled particularly. And they would have naturally the feeling that these damn scientists weren't practical men; they were visionaries. . . . And they didn't want to waste time on something that wasn't going to win the war." So back and forth the arguments went, with Loomis making urgent back door appeals to Stimson to do something. A week later, on June 19, Stimson wrote:

> Bundy has come back with word from Loomis and Karl Compton, who have been conducting investigations and experiments up in Boston. He reports them as saying the time has come to freeze the present situation—to waste no more time in experimentation but to go on and build plenty of instruments as we can with the knowledge we now have. They said the developments had gone along far enough so that we could depend on them now. There is always a reluctance of the Department to stop experimenting and I knew we would find it here particularly. . . . However, Loomis and Compton are going to be down here in person next Monday, so we arranged for another meeting with them to clinch the matter as to all the delay.

No matter how hard Loomis tried to push ahead, now regularly going over Bush's head straight to Stimson, he could not get the Army Signal Corps to move faster. On June 23, Stimson, after a conference on the delay in constructing airplane radar detectors, vented his own frustration in his diary:

It has been terribly held up by the finesse of the Signal Corps, who have been fussing over it for years instead of copying the workable arrangement the British have. I was fairly shocked to find how little they had done today. I dined with Bundy, and he had Loomis and Compton there, and also had in McCloy and [Robert] Lovett and we talked the whole thing over in the evening and if the fur doesn't fly tomorrow I'll miss my guess. The same old story of the better being the enemy of the good! and our Departments are worse sinners in this respect than anybody I know. They fuss over things trying to better them until the crisis is on us and the troops haven't got any of the equipment in question.

By keeping up the pressure, Loomis eventually achieved his end, and that spring Stimson ordered the first radar for the Army Signal Corps into immediate production. In the months to come, the white-haired secretary of war, who at seventy-three was in the position of having to evaluate and approve a whole new generation of advanced weaponry, would lean heavily on his cousin's technical expertise, as well as the scientific counsel of Bush, Compton, and Conant. Through these "dippings down," as Stimson called his practice of consulting directly with a trusted adviser on the progress in a specific field, he was able to cut through the bureaucratic double talk of the military and maintain a surprisingly accurate picture of what was really going on within his organization.

Meanwhile, Bush had been working on a solution to the stalemate. On June 28, Roosevelt, by executive order, created a new, greatly expanded organization called the Office of Scientific Research and Development (OSRD). Directed by Bush, with Conant as his number two, the OSRD would be run by scientists like Compton and Loomis as a flexible, fast-moving, and creative source of new weapons. They would be the first civilians to "push their heads into the generals' tent": they would be working toward military objectives, but independent of military control and unburdened by their outdated notions of what was and was not possible. The scientists had prevailed. Finally, substantial federal funds would be poured into university laboratories, not only greatly accelerating the pace of work, but enabling them to move beyond pure research to the production of revolutionary new devices that would make all the difference in the coming contest.

NOW that there was no longer any question that the laboratory would continue, it began to grow exponentially. By the spring of 1941, the Rad Lab staff had already grown to more than 140: 90 physicists and engineers; 45 mechanics, technicians, guards, and secretaries; and 6 Canadian guest scientists. Over that summer and fall, it would swell to almost 500 people, and more than $19 million would be committed to the secret radar systems they were developing and assembling. The penthouse roof laboratory was so dangerously overloaded that Cambridge authorities worried that it presented a serious fire hazard and urged MIT to relocate the whole operation to Mitchell Field on Long Island. Loomis and Compton dismissed this idea, but a new two-story building was slapped up and promptly filled to overflowing. More space on campus was procured, and almost every week, MIT students would arrive at a classroom only to find it sealed off and teaming with strange men.

Loomis and Lawrence's handpicked crew, which had labored with bunkered intensity on the AI radar, unencumbered by bureaucracy or, for that matter, any kind of formal routine, was evolving into a massive research and development organization. "It was a magnificent enterprise—staggering," recalled Bowles, notwithstanding his frequent complaints about Loomis' loose management style and indifference to housekeeping chores. "We went up by octaves on our money." He remembered being in Compton's office one afternoon when the MIT president was calculating that they had a budget of about $500,000 or so, and seeing where things were headed, he multiplied it by two. "But even then he was well under it," said Bowles. "There's nobody that can waste money like a physicist, but I think the result was extraordinary."

The sustained chaos of the first year could no longer serve as a management style, so an older, seasoned administrator named F. Wheeler Loomis (no relation), the longtime chairman of the Physics Department at the University of Illinois, was hired to sort out the personnel problems—not all egos adjusted equally well to teamwork—and impose discipline and order. A skilled bureaucrat, he was, as one early recruit observed, exactly what the unruly mob of prima donnas required, "a son of a bitch." If virtually every request for further funds or equipment had met with the approval of Alfred Loomis, under the day-to-day direction of the easygoing DuBridge, almost nothing got past Wheeler Loomis, who made frequent use of the word *no*.

As Loomis and his physicists kept envisioning new devices and setting off in new directions, the Rad Lab kept getting bigger and spawning new projects. The core group in the lab was still concentrated on Project I, perfecting AI equipment for aircraft. Lawson designed a new rugged spark-gap TR box, and in April it was incorporated into the B-18 bomber system, making it possible to pick up ships at a distance of fifteen miles. By late May, one of the rooftop model AI sets was sent at the army's request to Bell Labs for production, escorted there in the protective custody of two Rad Lab physicists.

At the same time the Roof Lab was mastering the art of ten-centimeter radar, Rabi, who was head of research, was already pushing on to the three-centimeter model, which would provide even sharper focus and more detailed information. As far as he was concerned, the lab's mission was "to develop something which could do as much harm as possible to the enemy." In considering any new tactical device, his standard query was "How many Germans will it kill?" Developing the three-centimeter cavity magnetrons demanded new components and even more challenging techniques, and while the military regarded it as another wasted effort, the policy makers at the Rad Lab were determined to pursue every promising avenue. Loomis made sure the three-centimeter project went forward, and it would succeed beyond all expectations. Because they now needed magnetrons in large quantities for their radar devices, Raytheon was also contracted to manufacture the disks and would supply the first three-centimeter magnetrons that would be used against the Germans.

Work on Project II, the microwave gun-laying radar, which had begun in January, was also progressing quickly. Loomis, in part because he was one of the few scientists at the lab with a background in astronomy, had suggested early on that they should use a conical scan to give precise azimuth and elevation, an innovation that played a vital part in the system's success. The other key role was played by Louis Ridenour, a brilliant and caustic physicist from the University of Pennsylvania, who bullied his group into going for broke in trying to build the first fully automatic tracking system. At the time, all naval fighting sets were manual, and automatic tracking was not considered feasible. Ridenour wanted to develop a ten-centimeter microwave radar set that could pick up an enemy plane on the screen, lock in on it, and follow it while continuously feeding the coordinates into a computer, which would point the antiaircraft gun at the target. After working out the theory for Loomis' conical scanning and borrowing freely from the physicists working on the airborne set, Ridenour's group was able to get a set to automatically track a plane from the roof of Building 6 on the last day of May. Six months later, an improved system was demonstrated for the Signal Corps at Fort Hancock. Obviously superior to its predecessors, it would become the prototype for the SCR-584 automatic tracking radar, one of the most important radar sets to come out of the Rad Lab, which was used by the army throughout the war. Thousands of SCR-584s would be deployed in battle and would play a crucial role in protecting the ground troops from air attacks.

THANKS to Loomis' preoccupation with what had come to be known as his "shower idea," the Rad Lab was also making great headway on Project III, the need for a long-range system of navigation independent of weather conditions. Back in October 1940, in the thick of the marathon planning sessions for the new radar lab, Loomis had talked to Bowen about Tizard's conviction that the North American continent was much better suited than war-torn Europe to develop and test a long-range navigation system. Loomis, "who must have been working a 24-hour day," recalled Bowen in his memoirs, "had fully appreciated this and practically overnight—on the basis of the description I had given him of the British GEE—came up with the suggestion of doing a similar thing. . . ." Pacing back and forth in the library of his New York penthouse, Loomis had excitedly elaborated on his idea to Bowen:

What about a pulsed hyperbolic system, like GEE, but on long waves which would be reflected from the ionosphere and would, therefore, give a range of one or two thousand miles. Since the two ground stations would themselves be about a thousand miles part, there was a problem of synchronisation. This he proposed solving using his specialty—in this case highly accurate quartz clocks—at each station.

Bowen had thought it a "marvelous idea," and from that time on, Project III had proceeded along the specific lines Loomis had suggested, becoming the basis for a new long-range navigation system, originally called LRN for Loomis radio navigation, though after Loomis objected to its being named after him, it was changed to Loran, for long-range navigation. Loomis had proposed a rather ingenious scheme in which pulsed radio waves from fixed shore stations would produce a grid of hyperbolic lines from which planes or ships, equipped with a specially designed pulse receiver, could fix their position. The key to Loran, as Alvarez later wrote, was Loomis' use of a time-measuring technique—a system of receiving and comparing the time of arrival of pulses—an expertise he had accrued during his long obsession with precision timekeeping:

> The Loran concept of a master station and two slave stations can be traced to the Shortt clocks, which had a master pendulum swinging in a vacant chamber, and a heavy-duty pendulum "slaved" to it, oscillating in the air. To obtain a navigational "fix" with Loran requires the measurement of the time difference in arrival of pulses from two pairs of transmitting stations. Each such time difference places the observer on a particular hyperbola. The observer's position is fixed by the intersection of two such hyperbolas, each derived from signals originating from a pair of long-wave transmitting stations. . . . The techniques for separating the signals and for measuring their differences in arrival time were "state of the art" at that time, but the problem of synchronizing the transmissions to within a microsecond, at points hundreds of miles apart, was a new one in radio engineering. Loomis proposed the following solution: the central station was to be the master station, and its transmissions were timed from a quartz crystal. The other stations also used quartz crystals, but in addition, monitored the arrival times of the pulses from the master station. When the operators noted that the arrival time of the master pulses was drifting from its correct value, relative to the transmitting time at that particular "slave station," the phase of the slave's quartz crystal oscillator was changed to bring the two stations back into proper synchronization. This procedure was able to bridge over periods when the signals at one station "faded out," and it was also what made Loran a practical system during World War II. . . .

With so many brilliant physicists pursuing independent lines of research, it certainly did not hurt that Loran was Loomis' own idea. His proposal was quickly approved by the microwave committee, and a group was set up to order the necessary parts, test equipment, and oversee the installation. A small group headed by Melville Eastham began work on the system early in the summer of 1941, and while waiting for equipment to arrive, they made a series of improvements, including moving to a longer wavelength to allow over-the-horizon operation. The basic system was completed in September, and the first field tests with a system using medium frequencies were conducted over the next three weeks.

A tunable receiver had been installed at Harvard's Cruft Laboratory, which had been made available, and another was set up at Lawrence's room at the MIT Graduate House. They had also obtained two abandoned lifeboat stations from the Coast Guard—one off Montauk Point, at the end of Long Island, and the other at Fenwick Island, off the coast of Delaware. They continued their investigation, running tests between the coastal stations and receiver stations in the Midwest in order to get a general idea about the behavior of sky waves over land. The main receiving station was set up by Donald Kerr in Ann Arbor, Michigan, in the home of the scientist S. A. Goudsmit, and they made control observations with a receiver mounted on a station wagon, stopping at Springfield, Missouri, and Frankfort, Kentucky. The tests strongly supported the possibility of stable sky-wave transmission and were so promising that they decided to abandon the original plan, which called for the ultrahigh frequency. As a result, Loran became the sole Rad Lab product not based on microwaves, an irony that was not lost on any of the Tuxedo Park pioneers.

Loran proved to be an extremely important new method of navigation, its principle virtues being that it was simple and highly successful. By means of Loran charts, created by the Hydrographic Office, an operator could plot his position accurately in about two minutes. More important, for wartime use, the ship or plane equipped with Loran emitted no signal that might betray its position to the enemy. It also proved relatively impervious to weather, with only severe electrical storms disrupting the system. By day, fixes could be obtained up to 700 miles from the transmitting stations, and by night, up to 1,400 miles. The NDRC immediately ordered that Loran be put into service in the North Atlantic. On September 25, Loomis reported on the Rad Lab's rapid progress to Stimson, who noted that it was "a very interesting talk. Things here at last seem to be jumping along."

While Loomis was congratulated for the dazzling ingenuity of Loran, there were those who found its similarity to the British system—the two

schemes turned out to be virtually identical, though at the time the
Americans were not permitted to know the details of GEE—too coincidental. Loomis' loyalists credited him with arriving at the idea independently, granting that the sketchy facts furnished by various members
of the Tizard Mission might have helped to "clarify or perhaps crystallize
the project." Bowles, who had always chafed at working under Loomis,
was outraged that the financier had somehow managed to usurp their
British partners, and he made no secret of it. His efforts to stir up controversy were stymied by Bowen, the mission's chief technical expert,
who fully supported Loomis' account of his bathtime brainstorm and
later testified to the fact when the navy applied for a patent in Loomis'
name.

Bowles could not let the matter drop. He was furious that there was
never any admission on Bowen's part that Loomis' "shower idea" was
anything but original: "Apparently, again, Alfred with his typical methods had been able to brainwash [Bowen]. He evidently captured Taffy
Bowen's fancy and in effect put Bowen in a position where he couldn't
do anything else but support Loomis' idea. In other words, it is my theory that Loomis would not have had the idea had he not been able to so
involve Bowen in a step-by-step process so as to find out exactly what
our British cousins had in hand."

It soon became obvious that Bowles and Loomis could not both endure under the same roof. "Loomis took a relatively possessive position
in respect to the radiation lab as if it were his baby," Bowles complained.
"I suppose with his ego and his past history, he wanted no competition."
He blamed the banker's frosty reserve for the lab's often difficult relations with outside agencies, particularly the navy, where Bowles had
good contacts. "His ways with the military were not the ways of a first-
class salesman. He worked with his cards too close to his chest, in fact
hidden in his vest when he wore one." Bowles made a clumsy attempt to
undermine Loomis' authority by criticizing him behind his back to
Compton, apparently unaware of their close friendship. When he lost
that argument, his bitterness increased. By the end of 1941, despite
Compton's efforts to alleviate the situation by assigning Bowles duties
that kept him at a safe distance, it became, in Bowles' own words, "an
impossible situation." Hoping for a showdown, he sent Compton a
memorandum marked "Personal" enumerating Loomis' grievous shortcomings as chairman of the microwave committee:

I have a few observations I wish to pass to you relative to the Microwave Section–Radiation Laboratory relationship which I hope will help your perspective of the problem. Please excuse the facetious tone—I am hurriedly doing this before I leave for Sperry.

First as to the Committee: We began operating as a Committee the first few meetings, then it became clear to me that Alfred did not want that kind of a Committee. The other Committees, as I understand them, have operated much as a unit. . . . Alfred's position is that the Committee members were merely directors to be informed and to be used when we went after funds. I believe I pretty well quote the sentiment expressed.

At the start Alfred told me that the two of us would have to look after things and I took him seriously. Later when I was removed from the Executive Committee running the Laboratory, I again understood from your memorandum on the subject that in a broad sense you, Alfred and I were to look after the general progress and policy. . . .

I felt at the time, and I spoke to you about it, that it would help my position administratively were I made Vice Chairman of the Section. As I remember it you seemed to approve the idea. I then spoke to Alfred and he brushed the idea aside by pointing to the strong position that of SECRETARY implied, the Secretary of War, the Secretary of Navy, etc. . . .

I suspect a certain minority in the Laboratory who do not like to recognize the authority of others have satisfied themselves that I am a "secretary" to write letters and do the bidding of others. This has made it hard for me especially when I have had to carry most of the administrative burden of the Section.

Alfred has done the job of a genius in so many ways, but I have had to try to keep a semblance of order into things and fill in those parts which did not interest him. . . .

Bowles' tone turned nastier as he went on to question Loomis' habit of ignoring policy, choosing instead to exercise sovereign authority over many of the lab's new projects and contracts: "This independence has been shown in the Laboratory's way of turning out reports of secret material without Committee approval in many cases; a procedure that I am sure is not the NDRC's. . . . It is the same independence that resulted in the Laboratory's giving Westinghouse a contract for ten-centimeter magnetrons without Committee knowledge. Perhaps Alfred did know. . . ." He concluded by assuring Compton that he was glad he "taking up whole matter with Van [Bush]," adding, "I give it only with the idea that it suggests a point of view among a *few*, perhaps only one or two militant ones—that may grow to bring embarrassment to you and the Institute."

Bowles no doubt realized how seriously he had underestimated Loomis when he was informed by Compton that his services at the laboratory were no longer required and that he was scheduled to take a position in the U.S. Army Air Corps communications area. "It was a polite way of banishing me," recalled Bowles. "Loomis had seen to it I was about to be sold down the river; in other words, his desire was to have me get the hell out, to use a common idiom. Compton sided with Loomis, if there was any side to be taken."

Bowles was saved from ignominy by Bush, who intervened at the last minute and, in an effort to avoid any more embarrassment than the imbroglio had already caused, persuaded Stimson's office to take him on as an "expert consultant" on radar. (Loomis had been the logical choice as the secretary's adviser on the new weapons, but concerns about nepotism forced Bush and Stimson to find another candidate.) Bowles left MIT for Washington in April 1942 and went on to become a highly effective ambassador for the new art of radar in Washington and played an important role in bringing the civilian scientists and the military high command closer together. But he never forgave Loomis, and years later he acknowledged that he found it infuriating that the former banker commanded so much loyalty and respect while he, one of MIT's original radar pioneers, had managed to win few friends among the close-knit Rad Lab crew. "An element resolved in our problems at the time was that I didn't belong to the fraternity of scientists who were brought in as the initial staff of the radiation lab. In other words I was not a physicist, I had no doctor's degree, and of all low brow things, I was an engineer," he said, adding bitterly, "I was a stranger in their midst."

Compton would always try to minimize the power struggle at the lab that resulted in Bowles' departure, and while praising the radar expert's abilities, he put it down to "the limitations of temperament and personality" that had led him into conflict with so many others in the past. Bush, who had his own rocky relations with Bowles dating back to the early years of his academic career at MIT, already knew about his ability to stir up strife. In the mid-1930s, when Compton had considered promoting Bush to vice president of MIT, Bowles had voiced an unfavorable opinion of the engineer, to the effect that he "had nothing but admiration for Bush's methods and not one damn bit of use for his methods." Bush, who had been promptly informed of the comment by Compton, asked Bowles to drop by his office, and the two men hashed out their differences for the next two and a half hours. But their relationship never recovered, and the two were often at odds. Bush always tolerated Bowles as a bright but "strange chap," too difficult and disloyal to be trusted. Bush always assumed Bowles' problems at the Rad Lab were largely of his own making: "It was Bowles against the field," Bush added, and "they pasted the hell out of him."

Bush, meanwhile, was not blind to Loomis' behavior and was fully aware of the adroit financier's propensity for masterminding events. There were times when their relationship became quite tense, particularly on those occasions when Loomis, in combination with Lawrence or the accommodating DuBridge, behaved as if the Rad Lad were an establishment virtually independent of the parent. An acutely skilled politician whose stern demeanor reminded some of a school principal, Bush was not shy about setting him straight. While he admired Loomis' energy and determination, he felt it necessary at times "to steer him to a path." According to Bowles, whose new perch in Washington occasionally afforded him the pleasure of seeing Loomis called in on the carpet, Bush "had a way of making clear in no uncertain terms who was boss. When dealing with a subtle plan or machination he was an artist in achieving a point by the use of memoranda concealing what was really on his mind."

But for the most part, Bush gave Loomis a great deal of latitude and made allowances for the fact that in civilian life he had not been accustomed to heeding the chain of command. For that matter, until he joined the OSRD, Bush was willing to bet that Loomis "had never taken an order from anybody at any time." As he observed years later, "Alfred's always been a close friend of mine, but a tough egg to work with. I think during the war when he occasionally changed his direction of action at my behest, it was about the only time that he ever paid attention to anybody over his head."

Chapter 11

MINISTER WITHOUT PORTFOLIO

Shall I just call up Mr. Roosevelt and ask him to save us all from a fate worse than death?

—WR, from "The Uranium Bomb"

DURING the long, frustrating spring of 1941, weeks that had been marked by heightening political tension, Loomis' involvement with the Rad Lab became increasingly sporadic as he found himself pressed into service on another scientific front. A number of leading physicists had become alarmed by the snail pace set by Lyman Briggs, the chairman of the uranium committee, to say nothing of the extreme difficulty in obtaining the materials and funds necessary for their experiments, and had brought their concerns to Loomis. The potential of the discoveries in nuclear fission had major implications, and scientists in both England and America were whispering about the possibility of constructing a bomb of enormous power. At this point, no one could reasonably doubt that America's involvement in the European war would increase, yet no one on the uranium committee could be persuaded that fission was critical to the war effort.

Worried that the Germans might already be ahead of them, they turned to Loomis, who had been instrumental in jump-starting the radar lab and was widely liked and respected, and begged him to use his influence to spur the government into action. Loomis disliked being drawn into these political squabbles, which he generally viewed as beneath him, but owing to his friendship with Lawrence, and pride in the achievements of the Rad Lab, which he justly felt would always extend to him, too, he began working behind the scenes to solve the uranium problem and push the government to get on with building the bomb.

In spite of the many obstacles, a year or more of research had yielded some important findings at various laboratories across the country, and Loomis shared his colleagues' doubts as to the Briggs committee's ability to appreciate the full weight of their implications. At Columbia, Fermi, working with Szilard, had gained a reasonably good understanding of the chain reaction in a uranium-graphite system, and Loomis knew they were in desperate need of more graphite to proceed with the next phase of their work. At the University of Chicago, Karl Compton's brother, Arthur, a Nobel Prize–winning physicist, was investigating beryllium as a moderator that could contribute to a successful chain reaction. Harold Urey, another Nobel laureate and the discoverer of heavy hydrogen, had done some promising studies showing it might be possible to obtain a chain reaction using heavy water. At the University of Virginia, Jesse Beams was working on isotope separation to achieve uranium 235, while at Harvard, George Kistiakowsky, who had worked for Loomis at Tower House and was now head of the university's Chemistry Department, was checking out gaseous diffusion as another possible means of separating uranium isotopes.

Meanwhile Lawrence, the great force in marshaling American physics, had become persuaded of the importance of nuclear weapons and was lending his voice to the growing discontent. Always on the lookout for ways to further expand his Berkeley operation, he was convinced that the Rad Lab should undertake uranium research, which was clearly receiving scant attention from the government. After all, his cyclotron had produced one of the major breakthroughs in the field. Back in the first flush of excitement surrounding the news of uranium fission, one of his boys—Ed McMillan—had devised an experiment to measure the energies of the fission fragments and in the process detected a mysterious new product of fission—a radioactive substance with a half-life of 2.3 days—which he speculated might be the isotope of the element

93. In subsequent experiments in the spring of 1940, he and another Rad Lab physicist, Philip Abelson, confirmed that it was the new element 93, derived by the capture of a neutron by uranium 238 and prompt subsequent decay.*

This discovery led McMillan to speculate that the element 93—which he suggested be named "neptunium" after the planet beyond Uranus, for which element 92 had been named—might decay to form an isotope of the element 94 with a mass of 239. After McMillan was drafted for radar work at MIT, and Abelson departed for Tuve's laboratory in Washington, their research was taken up by the Berkeley chemist Glenn Seaborg, who, aided by a young instructor, Joseph Kennedy, and a graduate student named Arthur Wahl, bombarded uranium with deuterons (the nuclei of heavy hydrogen atoms) and obtained a mixture of several isotopes of 93, including evidence of the element 94. Lawrence immediately suspected that 94 might prove fissionable—that it might have nuclear properties. If this was true, it could perhaps replace uranium 235 as a nuclear fuel or explosive.

In mid-December 1940, Lawrence, who was in New York staying with Loomis, met with Fermi at the office of Columbia University physicist George Pegram to discuss the feasibility of making enough of the new substance to do further research, which they all agreed was of the utmost importance. At this point, it was still all speculation. No convincing case could be made to the government without experimental proof of 94's fission characteristics. If the nuclear properties of the new isotope turned out to be unfavorable, the whole approach could come to nothing. During the meeting, Emilio Segrè, an Italian physicist who had worked with Fermi in Rome and was now a research assistant at Berkeley, suggested to Lawrence that they use one of his cyclotrons to manufacture enough 94 to measure its nuclear properties. "The only way of answering these momentous questions was by direct experimentation," Segrè wrote in his memoirs. "It was imperative to try."

After conferring with Loomis and Compton, and consulting Bush in Washington, Lawrence gave his assent. But he was already saddled with radar work and too weighed down by myriad projects and a heavy travel

---

* In nuclear chemistry, as explained by Arthur Compton in *Atomic Quest*, when a U-238 nucleus catches a neutron, its mass is increased by one unit, becoming U-239. But the nucleus is radioactive and emits an electron with a charge of minus 1 and of very small mass. The loss of one unit of negative charge increases the atomic number from 92 to 93, while the atomic weight remains unchanged at 239.

schedule to take on the uranium study himself. He had also been suffering from another of his debilitating bouts of flu and had spent that
Christmas convalescing in Florida. As it was, Loomis was so concerned
about his hard-driving friend's stamina that he proposed Berkeley establish a special fund to facilitate Lawrence's work for national defense.
Loomis felt Lawrence was much too valuable to the country to risk his
health and personally contributed $30,000 to the fund.

In the end, Lawrence decided to entrust the all-important experiments to Segrè, who would work on the slow neutron fission of elements
93 and 94 with Seaborg's crew. By this time, the British physicists at the
Cavendish Laboratory in Cambridge were becoming interested in 94
and wrote to Lawrence urging him to undertake personally the uranium
research. Events overtook the letter. Using Lawrence's sixty-inch cyclotron, Segrè and Seaborg's team of chemists had set to work immediately on bombarding uranium—this time with neutrons—and on the
night of February 23, they succeeded in identifying and producing samples of an isotope of the element 94. It had a mass of 239, just as McMillan and Abelson had predicted. It would prove to be the fissionable
isotope Pu-239—plutonium.

Realizing that the discovery of this new element and its transformation was of immense importance, Segrè informed Lawrence at once but
was not sure from his initial reaction to what extent Lawrence grasped
the full ramifications of their findings. "He told me to talk to his friend
Alfred Loomis, a multimillionaire banker and amateur physicist of great
intelligence who was visiting the Rad Lab," Segrè recalled. "I hesitated
because of security, but Lawrence reassured me that Loomis was cleared
for every technical secret concerning defense, and that furthermore he
was a cousin and close friend of Secretary of War Henry Stimson's."
Segrè reluctantly sought out the civilian banker, and this time he got
the reaction he had hoped for. "Loomis understood everything I told
him promptly and completely. I believe he helped to open Lawrence's
eyes, although it is possible that Lawrence had fully grasped what I told
him, and simply wanted Loomis to hear the news directly from the
horse's mouth."

Lawrence, who was shrewder than Segrè gave him credit for, recognized that his cyclotron opened up a whole new way of tapping nuclear
energy and had been planning to enlist Loomis' help in funding his plutonium research. He had sent Segrè ahead only to pique Loomis' interest, knowing his friend would take a keen interest in the project and

would be a powerful ally in pressing for a full-scale program. When Conant, Bush's deputy at the NDRC, came to Berkeley in early March
1941, Lawrence also went to work on him, urging him to "light a fire"
under the Briggs committee. "What if German scientists succeed in
making a nuclear bomb before we even investigate possibilities?" he demanded, asking the question he would repeat many times in the weeks
to come. Lawrence had been particularly dismayed by Briggs' reaction
to their work and felt his only real concern was with security. Briggs had
actually requested Lawrence "guarantee Segrè's reliability" and had
scolded him that Fermi had "only partial clearance."

ON a bitter cold New England morning on March 17, 1941,
Lawrence and Loomis met with Compton at his office at MIT.
Lawrence informed Compton that on his own initiative he had been
pursuing fission experiments, and although it was not technically his
NDRC assignment, he wanted to continue. He was sure that his results
merited further investigation, he told them excitedly, and it had occurred to him that he could convert his thirty-seven-inch cyclotron
into a huge mass spectrometer. Smaller instruments had already been
used to determine the mass and identity of isotopes. It would require
only minor modifications, and then it would be possible to use the mass
spectrometer to separate uranium 235 from ordinary uranium, thereby
isolating the fissionable isotopes. Lawrence, who could not keep from
leaping ahead, already envisioned a way the magnetic separation of isotopes could be expanded to become a large-scale process for producing
uranium 235.

Loomis and Compton needed no persuading and agreed to back
Lawrence's plan. That afternoon, Compton called Bush and recommended Lawrence be allowed to spearhead a new fission program and
cited the widespread dissatisfaction with the committee's delays and inaction. In a follow-up letter, he summarized Lawrence's harsh assessment of the existing fission program under Briggs, who was "by nature
slow, conservative, and methodical," and suggested that perhaps the
committee had accomplished so little because it "practically never
meets." This was particularly "disquieting," he added, as "our English
friends are apparently farther ahead than we are, despite the fact that we
have the most in number and the best in quality of the nuclear physicists of the world." Compton further recommended Bush appoint

Lawrence his deputy for ten days or so—just long enough to launch the program, using the same strategy that had served the radar program so well. Compton's tone was pointed: given the current crisis, they should be exploring all routes to the bomb, and the NDRC should not "passively administer" the uranium committee but "had a responsibility for insuring . . . that the project goes ahead not only safely but with the greatest expedition."

The truth was, Bush had his hands full and was charged with too many heavy responsibilities to adequately oversee the Briggs committee. But he did not appreciate being sandbagged by his own trusted lieutenants. He was not sure what irritated him more, that Lawrence had gone off on his own and come up with a method of separating the isotopes of uranium by what amounted to a mass spectroscopy stunt, or that once again he was being presented with an elaborate set of demands by Lawrence and Loomis, who as titans in the worlds of physics and finance had become a formidable duo. "Alfred and Ernest were great friends during the war and at times it got to be embarrassing," Bush recalled later, referring to their penchant for taking matters into their own hands. "Ernest had no sense of organization and he didn't have the slightest hesitancy in galloping right around me and going after the Secretary of War, the Congress or somebody. I don't think he ever tackled the President without my knowing it, but he would have been perfectly capable of doing it."

Of course, Bush often resorted to the very same tactics and frequently enlisted Loomis' aid in appealing directly to Stimson. The largest thorn in Bush's side was the difficulty in establishing and maintaining communication between the army and the scientists, and Loomis became a valuable go-between. Loomis was the only civilian to sit with a group of generals on an army planning board set up by the secretary of war to advise him on the V-1 and V-2 rockets being developed by the Germans. As Stimson noted in his diary, Loomis "looks very seriously on the rocket as a permanent change in military weapons, as important as the first use of the barrel for gun powder." It would be a result of that board's decision to take full advantage of all the modern techniques of warfare—including the SCR-584 developed in Loomis' laboratory, the proximity fuse being developed by Merle Tuve and his team, Bell Lab's latest advanced computer, and the army's antiaircraft guns—that the threat from the V-1s was substantially reduced. "Remember, Alfred had a very close personal relationship to Stimson, and that closeness was widely known, and it was something that Bush very much made use of," explained Caryl Haskins, who worked for Bush. Loomis' suite at the Wardman-Park Hotel in Washington, which he maintained throughout the war, served as a regular meeting place and was used by Lawrence, Fermi, Bowles, and other scientists when they needed to be available to talk to Bush or Stimson on very short notice. "Alfred spent a great deal of time in Washington administering the radar lab, and he was a force there," said Haskins. "He picked out some extraordinary people, like Kistiakowsky, and was of help on every technical issue, so that Van relied on him to a great degree."

Perhaps for this reason, Lawrence's back-channel lobbying of Loomis—along with Compton and Conant—particularly infuriated Bush, and he perceived it as a blatant attempt to undermine his authority. Lawrence had "decided that when he did not get directly out of me the reaction he wished, he would go around and bring pressure, which he certainly did," Bush fumed. When they met face-to-face on March 19, 1941, he let Lawrence have an earful: "I told him flatly that I was running the show, that we had established a procedure for handling it, that he could either conform to that as a member of the NDRC and put in his kicks through the internal mechanism, or he could be utterly on the outside and act as an individual in any way that he saw fit."

After the blowup, Bush moved swiftly to remedy the situation. Realizing Compton's advice made sense, Bush reluctantly named Lawrence Briggs "temporary personal consultant" and made funds available for further work on the elements 93 and 94. "I made such a nuisance of myself generally," wrote Lawrence, "that Bush requested the president of the National Academy [Frank Jewett] to appoint a committee to survey the uranium problem and make recommendations." Once Bush had deflected the political heat by appointing an independent review board, thus delegating the uranium problem to the country's top theoretical physicists and engineers, he and Lawrence patched up their differences. "I have not been at all disturbed in my own mind about the recent shindig," Bush wrote him. "There is no personality in the group that is not utterly reasonable when it comes down to brass tacks."

There remained the problem of deciding fission's future given the absence, as Bush saw it, of any "clear-cut path to defense results of great importance." Unfortunately, the review committee's first report, completed in mid-May, was not optimistic. They could not determine definitely whether an atomic bomb could be made. Arthur Compton, the committee chairman, concluded that "not a single member of the Briggs Committee really believed that uranium fission would become of critical importance in the war then in progress." There were still too many major scientific problems to be solved and too much confusion about how to approach fission. With no military application yet in sight, some committee members thought the entire uranium project should be shelved until peacetime, when it ultimately might prove a source of power.

In May, Loomis, accompanied by his wife, Ellen, went to Berkeley to collect an honorary degree from the University of California for his work in physics and for "devoting himself single-mindedly to the defense of our country in a time of emergency." Instead of celebrating, he and Lawrence spent nearly all their time plotting what steps they should take next to spark-plug Compton's committee, of which Lawrence was a member. In the few hasty letters they exchanged that summer, they continued to keep each other informed of any progress in what they referred to as "the U matter."

For much of the summer of 1941, Loomis was busy with radar work, and Lawrence was distracted by other demands on his time. The NDRC had asked him to go to San Diego to help stimulate the anti-submarine program, as the Germans were threatening to destroy the convoy system and Britain was on the brink of collapse. Months of relentless U-boat attacks had claimed 328 merchant ships, sinking tons of desperately needed supplies—wheat, beef, butter, explosives, and oil, as well as military equipment. Once again, Lawrence acted as the chief recruiting agent for the new underwater sound laboratories and rounded up former protégés from around the country. He even summoned McMillan from the radar lab to help him out, much to Loomis' dismay. As Lawrence wrote in a July 10 letter to Bush: "Alfred and Karl, I'm afraid, have a little feeling I'm walking out on them in being so concerned with the submarine program . . . [it] seems so clear the microwave committee is well along . . . on the other hand the submarine program has not gelled and there is urgent need for the best scientific talent."

But by fall, Lawrence was back at Berkeley and was convinced that every effort should be made to build a bomb using either uranium 235 or plutonium 239. Over the summer, he had learned that John Cockcroft, using the cyclotron at Cavendish Laboratories, had obtained uranium 235 and had persuaded the British uranium committee of its merit in a "superexplosive bomb." Moreover, Mark Oliphant, another of Britain's radar pioneers, was now confident that a U-235 bomb was within the realm of possibility. During a visit to the Rad Lab late that autumn, he told Lawrence of the growing conviction among British scientists that if he tackled the uranium problem, real progress could be made. Lawrence promised Oliphant he would do what he could.

True to his word, early that September Lawrence called Arthur Compton and told him that they had to jump-start an atomic research program—there was no time to lose if America was to win the race for the bomb. "What he was most certain of," recalled Compton, was "that plutonium 239 could be made in a chain reacting pile and then separated by chemical methods from its parent, uranium 238. His evidence was good. . . ." There was nothing really new in Lawrence's argument, but it was his absolute confidence in fission as a concrete military weapon, and unshakable faith that it held the key to victory, that gave the program new life. As Compton wrote in his memoir, *Atomic Quest*, Lawrence's unique contribution "was a feasible proposal for making a bomb. No one else ever proposed the possibility. He came forward with what he felt could be carried through, and had something tangible to take hold of."

Not long after that phone call, Lawrence, Conant, and Arthur Compton gathered before the fireplace in Compton's Chicago home. Lawrence outlined the Berkeley experiments that indicated a bomb could be made with only a few kilograms of fissionable material, either uranium 235 or plutonium. He described the various ways uranium separation could be achieved and argued in favor of pursuing several lines of investigation simultaneously. Every effort should be made to further Fermi's pile experimentation and plutonium production at Columbia, while he continued testing the more expensive technique, magnetic separation, using the mass spectrometer at Berkeley. Conant, who like Bush tended to err on the side of caution when it came to fission, weighed in on the need to concentrate on projects that were certain to be useful and the importance of not wasting their resources on such an unproved assumption. But after an impassioned speech by Lawrence, assisted by Compton's reports of Nazi efforts already under way, Conant came around.

Aware of Lawrence's legendary determination to build a colossal 184-inch cyclotron, Conant was somewhat skeptical of the Berkeley physicist's assurances that he would commit his laboratory to the effort. Not one to mince words, Conant asked Lawrence point-blank if he was prepared to devote the next several years of his life to the atomic program. The question "brought up Lawrence with a start," recalled Compton, who never forgot the expression in Ernest's eyes as he sat there, "his mouth half open." Lawrence hesitated only a moment—" 'If you tell me this is my job, I'll do it.' " With those words, he signaled the start of the race for the bomb.

Typically, Lawrence, with a "damn the torpedoes" enthusiasm even Admiral Farrugut would have admired, threw himself into the uranium work. He gave the project priority over everything else, quickly raising money from Loomis and other longtime patrons. He insisted on bringing in Robert Oppenheimer, a close friend and Berkeley colleague, who was an outstanding theoretical physicist. The only problem was that Oppenheimer was regarded with some suspicion for his union ties—he had tried to recruit faculty members for the American Association of Scientific Workers—and Lawrence had cautioned him more than once to stop his "leftwandering activities." But Lawrence, who had great respect for "Oppie," was insistent: "I have a great deal of confidence in Oppie," he wrote Compton, "and I'm anxious to have his judgment in our deliberations." His clearance was soon authorized. On October 21, Loomis and Lawrence attended a meeting of the uranium committee—now designated somewhat cryptically as the S-1 Section of the newly formed OSRD—at the General Electric research laboratory in Schenectady. The purpose of the meeting was to discuss how destructive the bomb would be and how much time and money would be required to build it. Oppenheimer estimated that it would take a hundred kilograms of U-235 for an effective explosion. None of Bush's "hardheaded" engineers on the committee dared even hazard a guess as to the number of years or total cost—somewhere in the hundreds of millions of dollars—involved in building such a weapon.

While the committee struggled to answer those questions to Bush's satisfaction, Lawrence got to work on modifying and enlarging the thirty-seven-inch cyclotron. On November 6, Arthur Compton personally handed Bush the committee's report, complete with the calculations by theoretical physicists such as Oppenheimer and Kistiakowsky, acknowledging that "a fission bomb of superlatively destructive power will result from bringing quickly together a sufficient mass of element U-235." Their recommendation: A full effort to make atomic bombs was essential to the safety of the nation. Roosevelt acted at once and formed a high-level committee to consider the fission bomb proposal, appointing Bush, Conant, Stimson, General Marshall, and Vice President Henry Wallace. Exactly one month later, on December 6, Bush delivered the president's verdict: If atomic bombs could be made, America had to make them first. Although no record of Roosevelt's words exists, he made it clear to Bush that he wanted work on the atomic bomb project to be "expedited . . . in every way possible."

That same day, Lawrence's newly completed thirty-seven-inch mass spectrograph was used to separate a small amount of U-235, and as he happily reported to Loomis, he was "pushing with the boys" on the isotope separation problem:

> I am glad to report that we have already made excellent progress. We have had as large a beam of ions as I anticipated we would be able to produce, i.e. enough to get about a microgram per hour, which is about a hundred times more than others have obtained. We are making some further modifications of the mass spectrograph, and within a few days I hope we will be grinding out a substantial sample.
>
> There is little else to report.
>
> Molly and I privately cherish the hope that we may be able to be with you at Hilton Head [for Christmas]; at least it is pleasant to contemplate the possibility.

On Sunday morning, December 7, 1941, the Japanese hit Pearl Harbor. The following day, December 8, Congress declared war on Japan. The attack immediately unified the country and silenced the isolationists and fission naysayers. Now everyone was focused on a common goal—beating the Nazis to the bomb. America was now at war on two fronts and could not allow the damage done to the country's pride and naval strength to give hope to their enemies. In the weeks and months after the Pearl Harbor attack, Lawrence worked night and day on perfecting the mass spectrometer to obtain the precious U-235. As he wrote Loomis on December 12, "Now that the war is on I am more anxious than ever to be useful."

As was by now his habit, he discussed every detail with Loomis, analyzing the various approaches and worrying that time was "too short." Lawrence was convinced that with hard work and persistence he could overcome any obstacle or find a way around it by means of a new invention, and he was insistent that every potential method be explored, that no possibility be overlooked—"Let's try them all," he would tell Loomis as they hunched over his notebooks. He was frustrated at the thought of how much time had been wasted on policy debates, and as a consequence they now found themselves in a breakneck race against German science. "The one thing he wanted to get was to separate U-235," recalled Loomis, referring to the many technical problems involved in the electromagnetic separation of uranium. "He invented the 'calutron' [an improved method of magnetic separation] when he was sitting in my apartment in New York. He was making sketches." Lawrence took out a sketch to show him that he had made a week or two earlier when he suddenly said, " 'I've got a better way,' " which, Loomis added, "turned out to be true."

Their baby—the 184-inch cyclotron—which was under construction in a new hilltop laboratory high above the Berkeley campus, would be converted into a large-scale spectrograph, or "calutron." The big magnet was erected first and a building hastily constructed around it. It reminded Loomis of his early, heady days of cyclotron electronics, and if it were not for his responsibilities back at the Rad Lab, he would have loved to stay and be part of the excitement. In May 1942, the huge spectrograph was turned on for the first time, but it was many months still before it was successfully separating the isotopes. Lawrence's cyclotroneers worked 365 days a year for the next three years to invent around the difficulties and make the calutron work. His fantastic mass spectrograph would become the basis of hundreds of other calutrons soon to be built and became the main method of producing uranium 235 for the first atomic bomb. Arthur Compton headed a parallel program to breed plutonium in quantity at the Metallurgical Laboratory in Chicago. Fermi would run the newly formed Argonne Laboratory, testing the graphite bars and uranium slugs to be used in the reactors. Everyone who was available pitched in to help, and it is safe to say there was not a single unemployed physicist in the country.

To make enough uranium and plutonium fast enough required a herculean effort. Using his uncommon talents as a leader and indomitable energy, Lawrence began the process of staffing a mass spectrometer plant in Oak Ridge, Tennessee, hiring bright young graduate students left and right and once again robbing Loomis' Rad Lab of skilled men. In only a short time of frenzied activity, he managed to build up a large-scale operation. Lawrence's long-standing friendship with Loomis allowed for an unprecedented and continuous exchange of talent and ideas between their two scientific divisions, particularly as many of the physicists at MIT were drafted to work at Los Alamos. "The fact that Lawrence was a member of this radar section aided in maintaining the cooperation between the radar and nuclear programs," wrote Arthur Compton. "Such cooperation became essential to the completion of atomic weapons."

As Alvarez reflected years later, it was a great stroke of luck for the country that Loomis was involved in the uranium project from the beginning, not as an originator of ideas so much as an individual who knew how to exploit them, and his actions would contribute to "the remarkable lack of administrative roadblocks experienced by the Army's Manhattan District, the builders of the atomic bombs." This "smooth sailing," he wrote, "was due in large part to mutual trust and respect the Secretary of War and Alfred had. Alfred was in effect Stimson's minister without portfolio to the scientific leadership of the Manhattan District—his old friends Ernest Lawrence, Arthur Compton, Enrico Fermi, and Robert Oppenheimer."

Chapter 12

LAST OF THE GREAT AMATEURS

Ward's expression had not changed; he was very pale and there
were blue circles under his eyes.

—WR, from *Brain Waves and Death*

LOOMIS' work had placed him at the very center of the atomic
bomb debate, but in the wake of the disaster at Pearl Harbor, he had little time for Lawrence and his uranium separation project. The terrible
and unexpected defeat showed just how poor the nation's radar defenses
were against an air attack. In the confused and bitter aftermath, Washington was awash in guilt, and half a dozen boards were quickly set up,
busying themselves with assigning responsibility for the catastrophe to
various military departments and officers. Like most Americans,
Loomis was shocked by the overwhelming reports of the carnage, which
had killed more than 2,400 servicemen and civilians, wrecked eight
battleships, three destroyers, and three cruisers, and left the better part
of three hundred aircraft in smoldering ruins. He had felt it all the more
keenly because his youngest boy, Henry, a navy ensign, was stationed in
Hawaii on the *Pennsylvania*, and it was several days before he heard that

he was safe. But the blow also demonstrated what he had been arguing for months, that the radar systems being developed at his laboratory were vital to their ability to protect themselves against an attack on American territory. Even on December 7, Loomis knew that the months of indecision and relative inaction had finally come to an end. He was no longer exploring the speculative art of radar for possible defense applications, he was in the business of building detection devices for the offense.

On December 10, three days after the strike on Pearl Harbor, the Japanese invaded the Philippines and destroyed Britain's brand-new battleship *Prince of Wales* and the famous old battle cruiser *Repulse*, along with their escort of destroyers. The Japanese launched an amphibious offensive against Malaya, overrunning the British defense with their superiority in the water and the air, and were driving south to Singapore. The following day, Germany and Italy declared war on the United States. America was in a shooting war on two fronts, and it was clear that radar—airborne and shipboard—would play a leading role in the conflict. At this point, Loomis' Radiation Laboratory had been in operation for little over a year and had completed most of the basic research, achieved the important breakthroughs, and carried out field tests. It had an impressive array of prototype systems in operation. But because of the skeptics in the military, and entrenched resistance to new techniques and experimental models, not a single ten-centimeter microwave radar system was in use in combat anywhere. In one stroke, all that had changed. Just as the Chain Home system had played a decisive role in the Battle of Britain, Loomis knew with certainty that the Rad Lab's powerful radar equipment would be the critical factor in the Allies' favor.

Among the first systems to be put into service was an RCA production model of an air-to-surface-vessel radar that was salvaged from the wreckage of the USS *California* in Pearl Harbor and quickly set up at the Oahu radar training school, where Loomis' son served as an instructor. By December 1941 the Rad Lab ASV radar was detecting ships twenty to thirty miles away and locating submarines at a distance of two to five miles. Thanks to the invention of the circular oscilloscope screen called the "plan position indicator" (PPI), pilots could determine the vessels' location at a glance and calculate their range and bearing relative to their plane. The U.S. Navy had been sufficiently impressed to order ten experimental ASV sets for their own pilots and had ordered a hundred shipboard microwave search units for their destroyers. The British had been so eager to equip their pilots with the microwave radar device that they had diverted two of their B-24 bombers to be fitted with the ASV sub-hunting system, subsequently dubbed "Dumbo I" because the radar dome in the swollen nose of the aircraft gave it an elephantine look. On December 11, as America went to war in Europe, Dumbo I made its first test flight. That same week, the Rad Lab physicists began converting all the available AI sets into improvised ASV search radar systems to send out after German subs close to America's shores.

In those first weeks after the United States declared war, while the army raced to install the Rad Lab's modified microwave radar sets in their B-18s, the country paid a high price for its lack of preparedness. The Allied strategy, agreed upon immediately after the United States officially entered the conflict, was to defeat Germany first and then go after Japan. German U-boats already controlled the waters off the eastern coast and the Gulf of Mexico and were inflicting devastating losses: in February alone, eighty-two U.S. merchant ships were sunk by Nazi submarines. Almost every day of the war brought news of the sinking of two or three more tankers, and their steel carcasses littered the Atlantic floor. Without the aid of radar, army and navy pilots managed to attack only four Nazi submarines in the first two months of patrol.

On April Fools' Day, the first B-18 with the modified ASV radar search units took off from Langley Field. On its first night patrol, it spotted three U-boats. It gave chase and, zeroing in on the enemy at a range of eleven miles and an altitude of three hundred feet, found and sank one of the submarines.

From then on, America's scores improved steadily. A few days later, the USS *Augusta*, armed with the first production model of shipboard search radar, joined the fight. All that summer, the roaming eye of the Rad Lab ASV radar had the German wolf packs on the run. The U-boats, equipped with receivers that had been designed to pick up the old long-wave ASV, were not capable of detecting the microwave pulses from the Allies' new search radars. By late summer, convoy losses dropped sharply. In Berlin, German admiral Karl Dönitz, who in 1940 had boasted that "the U-boat alone can win this war," was forced to admit that "the methods of radio-location that the Allies have introduced have conquered the U-boat menace." As he later wrote, radar threatened to provide the Allies with the key to victory unless Germany could address the disparity in their technological prowess:

For some months past, the enemy has rendered the U-boat ineffective. He has achieved this objective, not through superior tactics or strategy, but through his superiority in the field of science; this finds its expression in the modern battle weapon—detection. By this means he has torn our sole offensive weapon in the war against the Anglo-Saxons from our hands.

Ironically, Loomis was helped in his campaign to sell radar to the military's top brass by his old nemesis, Ed Bowles, who was now an expert adviser to the secretary of war. Loomis had consistently sought to keep Stimson informed of the advances in the Rad Lab radar and hoped that he might use his influence on the newly formed Joint Committee on New Weapons and Equipment (JCNWE), a panel that had been formed under the Joint Chiefs of Staff to coordinate civilian research for the war effort. Loomis' efforts were not wasted, for Stimson was so unhappy about the services' conservative approach to the new technology that after personally checking out a demonstration of the ASV-equipped Dumbo II, which successfully located a distant ship, he fired off irate notes to Generals Marshall and Arnold demanding action: "I've seen the new radar equipment. Why haven't you?" But it was Bowles, eager to establish a role for himself in Washington, who wrote persuasive reports for the JCNWE, courted the various generals, and assiduously worked from within to change their perception of microwave radar. He pointed out the ways in which the new weapons were not being deployed to their full potential, arguing that the Rad Lab ASV could not be treated "simply as a magic gadget" but needed to be part of "an operational framework." By August 1942, Bowles—who according to Bush had virtually become assistant secretary of war for radar—was winning support for the establishment of a regular ASV-equipped bomber unit at Langley Field to search out and attack German submarines, and for a new offensive strategy—air search—that would do so much to finally win the U-boat war.

WITH the demonstrable success of the ASV in the Atlantic, and new appreciation for the strategic opportunities presented by the advancing technology, there was a mad rush of activity at Loomis' laboratory. The Allied air forces were revving up for a full-scale bombing offensive against continental Europe, and the physicists went to work on advanced radar aids to bombing. There was a feverish drive to construct twenty of the Rad Lab's highly accurate three-centimeter sets, code-named H2X, so that they could be shipped to England before the cloudy fall weather set in. The effort to develop radar beacons, echo amplifiers to be used as blind-landing aids, was reorganized and accelerated. By the summer of 1942, the CXBK, an experimental microwave ASV radar, was in operation over the Bay of Biscay, and when the British reviewed pictures of its scope presentations, with their clearly defined coastlines and bays, the towns and cities standing out brightly, an RAF dignitary declared, "Gentlemen, this is a turning point in the war."

The Rad Lab physicists scrambled to meet all the requests for radar equipment, took on dozens of new projects, and came up with still more innovative gadgets. Most important, they finally overcame the wariness of the military services so that they could work closely together in developing new tactical devices and guarantee that they would be successful in the field. Loomis, who was a strong believer in the necessity of a "follow through" policy, fought to increase the Rad Lab's role, declaring that it should not only stay in the engineering business, but should be expanded greatly to handle more of it. His microwave committee recommended "a several-fold increase in the number of scientists and engineers engaged in its research and development program."

Loomis' drive to increase the size of the lab was stoutly opposed by a number of leading industry representatives, who jealously accused Loomis of creating his own private factory and leveled charges of government encroachment. But with Compton's backing, and the unanimous support of the microwave committee, Loomis' proposal carried the day. The Rad Lab was now permitted to create a "model shop" to produce limited numbers of the experimental microwave radar it was designing for the military, and given unprecedented freedom to operate, it became more a partner to the army and navy than a mere adviser to manufacturers. At the outset, Loomis had worked hard to achieve a "meeting of minds" with industry, but after a while, as Bush put it, "The Radiation Laboratory took the ball and ran away." Some of the fights between the sides became quite entrenched. The big companies were "damn conceited," in Bush's view; then again, the physicists were "also conceited." Much of the time he felt like saying, "A plague on both your houses."

"Loomis thought that we were fighting the war and not doing pure research, that our job was not only to develop the equipment but to get it into use," explained DuBridge. "There were people who felt that we put too much effort in the field and there were other people who felt we didn't put in anywheres near enough. We tried to strike a balance," he said, but "the idea that we would see our equipment help win the war was basic of the laboratory philosophy from the day we began."

Still, convincing the military to cooperate with the scientists was not easy. Rabi, who was in charge of advanced development, vividly recalled what happened when one navy contingent came to the lab to describe the various devices they wanted: "I asked, 'What are they for? What is their purpose?' This naval officer looked me in the eye and said, 'We prefer to talk about that in our swivel chairs in Washington.' " Rabi did not answer, but he did not do anything, either. Engineering specifications alone would not help foresee combat needs. When they returned again with another problem, he told them, "Now look, you bring back your man who understands radar, you bring your man who understands the navy, who understands aircraft, you bring your man who understands tactics, and then we'll talk about your needs." Taking that from one of the "longhairs"—the academics—"was pretty hard for them to swallow," acknowledged Rabi, "but they did." From that day on, they started working together as a team:

> They were worried about snooper planes following the fleet, and they wanted to shoot them down. We developed a height-finding radar to be used on the ship, a radar to be used in conjunction with airborne radar. This started a relationship with the navy that was very important to us. When we got to know one another, when they learned we were trying to help them and that we respected them, when they discovered we didn't want any of the glory, we came to be friends with great mutual respect.

The Rad Lab scientists no longer confined themselves to their cubbyholes but often developed their weapons in the field, traveling to the European and Pacific theaters of operation to assess the military's needs, going back to the lab to fine-tune their design, and then returning to the front to oversee the deployment of their devices.

Because the Rad Lab could not accommodate all the additional projects they had taken on at the request of the army and navy, Loomis arranged for additional funds from the NDRC, and the laboratory began to grow rapidly. He first purchased the Hood Milk Company building on Massachusetts Avenue, just two blocks away, and had the three-story brick structure overhauled as laboratory space, and then he bought the neighboring Whittemore Building. Four floors and a penthouse were piled onto one of the original labs, and field stations cropped up in Orlando, Florida, Spraycliff, Rhode Island, and Deer Island in Boston Harbor. New men were brought in and trained to run these facilities, and an army of young women came—and kept on coming—to meet the lab's growing administrative needs, working as secretaries, bookkeepers, and technicians, so that they eventually outnumbered the men almost two to one. By the end of the year, the staff would reach almost two thousand people and was still growing, with a budget of $1,150,000 monthly. Loomis' small secret community of physicists had evolved into a large, energetic, affectionate mob, complete with raucous parties—comings and goings were vigorously feted—romances, and intralab marriages. Any concerns that were raised about the reproductive effects of all the high-frequency energy emanating from the early radar sets were put to rest when three people in the aptly named "propagation group" became parents at the same time.

Under Loomis, the Rad Lab moved from research into development, design, and manufacturing and was reorganized into ten divisions working in parallel and reporting to a steering committee. Dozens of new departments were created to handle the great flood of orders. In the early days, with everyone acutely aware the Germans were pounding American ships, everything was done with an eye to speeding the lab's progress: the personnel office expanded to organize the tremendous human traffic; the accounting office kept the books and paid the bills; the business office handled the maze of requisitions, purchase orders, and invoices; the purchasing group bought the thousands of mechanical and electrical hardware items—the largest item being a carnival merry-go-round—required by the physicists; the self-service stockroom carried over five thousand electronic parts; the receiving room handled, at its peak, a monthly average of twenty thousand boxes; the mailroom, which by the end employed twenty-five very popular girls, sorted the sacks of internal and outgoing letters and documents; and the shipping room moved the tons of radar equipment that had to be crated and loaded before leaving for the docks and England. After a luncheon meeting with Loomis that fall, Stimson noted in his diary: "His work in Boston seems to be progressing satisfactorily. His inventions are making rather more rapid progress than we can take care of them."

The Rad Lab had officially entered "the big time," with a total of fifty projects on its books. The lights burned all night, and the scientists, who lived on coffee, doughnuts, and sandwiches, worked twelve hours a day, seven days a week, that first year. "Tea wagons" loaded with equipment constantly rattled down the corridors, and as the yearbook notes, "Things got so busy it was fashionable not to answer your phone." Security was tightened, identification badges were mandatory, and despite the well-intentioned warnings about secrecy, gossip remained the major form of recreation. Everybody talked all the time—whether it was about the imminent danger of enemy agents or the latest dope on a newly launched project. Despite the constant hum of conversation, there was never any evidence then or since of a single laboratory leak. The Rad Lab was its own world, self-contained, urgent, and alive, and it absorbed everything so that the outside world seemed almost distant. Alvarez, who had been sidelined for several months by surgery and serious side effects from the anesthesia, returned to a laboratory that was very different from the one he had left. "We really were working as industrial scientists now," he recalled. "Activities and staff had grown exponentially. It took me a while to catch up."

One of the projects Alvarez was originally assigned to—the automatic tracking fire-control radar—had been completed in his absence, and an operating SCR-584 prototype, the first of its kind, had been mounted on the MIT roof. The second, known as the XT-1, which was mounted on a truck, had been completed a month before Pearl Harbor and sent to Fort Monroe in Virginia for evaluation. Alvarez, who became something of a Rad Lab legend for the three words *Suppose we have*, often found scrawled in his laboratory notebooks, had another idea. One day back in August, watching the first microwave fire-control radar automatically track an airplane, he suddenly realized that if a radar set could continuously track an enemy plane accurately enough to shoot it down, it should be able to use the same information to guide a friendly pilot to a safe landing in bad weather. With strong support from Loomis, Alvarez had begun developing the concept that would emerge as one of the most valuable contributions of the Rad Lab—an aircraft blind-landing system, or "ground-controlled approach" (GCA). Thanks to the accelerated post–Pearl Harbor development schedule,

XT-1 prototypes became available sooner than expected, and early in the spring of 1942, Alvarez and a group of about twenty physicists got down to work testing his idea.

They had to prove first of all that it was possible to "talk" a pilot down—that he could land solely on verbal instructions from the ground. In the Logan National Guard hangar, now completely taken over by Rad Lab projects, they had a blindfolded pilot walk a line on the floor while a controller observed his deviation and instructed him how to correct his path. The results were encouraging, so they designed further tests, rigging a primitive radar set with optical sights. Alvarez also took lessons at the Squantum Naval Air Station several times a week to prepare for the navy instrument pilot's exam, on the theory that he would better understand the challenge pilots faced in flying on their instruments alone and perhaps eventually learn to land by GCA himself. By late March, Alvarez and his small team drove to Quonset Point Naval Air Station outside Providence and began conducting trials of their makeshift blind-landing system. During several weeks of work with a navy test pilot named Bruce Griffin, Alvarez gained confidence in their "unorthodox" solution to the blind-landing problem:

> We improvised communications procedures as we went along, and after many landings under conditions of good visibility Bruce began to fly under the hood [with the hood literally over his head to prevent sneaking a peak] to lower and lower elevations before making a visual approach. His radioman acted as check pilot when he was under the hood. One day Bruce announced that he had flown under the hood all the way to touchdown. We cheered the first complete landing on GCA.

In May, the navy invited them to test their radar system at the Oceania Naval Air Station near Norfolk, Virginia, offering an advanced SNJ trainer, considered by pilots to be one of the best planes ever built. "It was ideal for the aerobatics," recalled Alvarez, who went along for the exciting ride as "Bruce brushed up on the loops, rolls, split-S turns, and Immelmans [reverse turns] he couldn't attempt in his ungainly Duck." Unfortunately, the XT-1, which had worked so beautifully in its one previous test tracking a low-flying aircraft, went completely "spastic." Every time the plane approached the truck, the antenna would suddenly break away from the line of sight to the plane and point at its reflected image three degrees below the surface of the runway. They tried every trick they knew to keep the radar from locking in on the reflected image, but the equipment was unable to track planes near the ground. The GCA tests were "disastrous," and Alvarez and his team returned to MIT thoroughly discouraged. He knew the most sensible option was "to give up and move on." There were many other important radar problems that needed to be solved, and they were wasting manpower and resources. But they had "fallen in love with the GCA talk-down technique," and it was the only hope they had of solving the problem of landing planes in fog and poor visibility conditions, which regularly grounded bombing and air reconnaissance missions.

The crucial breakthrough came during a dinner with Loomis at his suite at the Ritz-Carlton in early June. Alvarez, who by then was feeling defeated, gave Loomis a frank assessment of the system's shortcomings, and the two of them then analyzed the Quonset Point and Oceania events in detail. Loomis was obstinately committed to the idea that GCA was feasible and was ready to throw out the conventional ideas of what was possible in the existing art of radar. He was insistent and, as Alvarez later recalled, "did an amazing job of restoring my morale, which had hit a new low." He also made it perfectly clear that neither of them was leaving until they had hit on some sort of solution. "We both know that GCA is the only way planes will be blind-landed in this war," Loomis told the thirty-one-year-old physicist, "so we have to find some way to make it work. I don't want you to go home tonight until we're both satisfied that you've come up with a design that will do the job." Together, they engineered a radical new form of radar, as Alvarez recounted in his memoirs:

> We both contributed ideas. The antenna configurations we devised departed completely from all previous designs. Out of the Ritz-Carlton discussion came a tall, narrow, vertical antenna that scanned by being mechanically rocked and a horizontal antenna that scanned its pattern left and right. The beaver-tail beam from the vertical antenna would be so narrow that when its main lobe pointed at a plane very little energy would spill even one degree away, eliminating the possibility that the system would confuse the plane with its reflections. Alfred suggested switching a single radar transmitter between the two antennas four times a scan cycle. The three principles Alfred and I came up with on that memorable evening have been basic to

GCA technology ever since. I was allowed to leave for home just before midnight.

Had it not been for Loomis' challenge that night, there might not have been a blind-landing system in World War II. "I would have immersed myself in other interesting projects to forget my disappointment and embarrassment," wrote Alvarez. "Many lives would have been lost unnecessarily."

Once again, the fact that Loomis had played a key role in the conception of the new GCA design had its advantages. He would often drop by Building 22 to offer encouragement, and after Alvarez and his graduate student Larry Johnston had worked out the kinks in the design, he urged them to build a demonstration model as quickly as possible, promising to use OSRD funds to have ten prototypes built by a small radio company on the West Coast. Loomis, who had honed his competitive skills on Wall Street, felt that farming out the contract to a small firm would yield both faster and better results, and as usual, he was hedging his bets by engaging in a little frontal maneuvering. He had for some time now been concerned about the wide gap between the physicists' technological innovation and its effective production. Loomis wanted to see GCA succeed and was troubled by the bad case of "NIH" (not invented here) that the industrial engineers had developed after one of the Rad Lab's previous airborne sets had failed to translate well in manufacturing. The "reengineered" model that they had finally delivered months late was so cumbersome, it never saw any action.

Determined to avoid a repeat of this scenario, Loomis, together with Rowan Gaither, a very able San Francisco attorney who went on to become a close friend and business partner, created the transition office—known informally as the "hurry-up department"—to shepherd the lab's creations through manufacturing and production and, in many cases, right into the field. The crux of the problem was that microwave radar was still so novel, few manufacturers had adequate facilities and trained personnel to produce the devices. Loomis, who always strove to keep red tape to a minimum and money easy, appointed Gaither as chief overseer and troubleshooter. To keep production problems to a minimum, Gaither would invite industrial engineers at companies of their choosing to come to the Rad Lab and receive training in the intricacies of the physicists' creations and proper testing of the new equipment. In short, they would be educated in the Rad Lab way—which roughly translated as "*our* way or the highway." This basic strategy proved such a success, it became a matter of laboratory policy to thus facilitate the transition from prototype to production model, much to the irritation of several industrialists, who again groused about Loomis' unfair tactics.

Loomis' other reason for ordering the ten preproduction models of the embryonic devices, designated the Mark I, was to circumvent objections from the army and navy, as well as the RAF. According to Alvarez, they had all let it be known that their pilots would "never obey landing instructions from someone sitting in comfort on the ground," and they preferred to wait for something along the lines of the ILS (instrument landing system). Loomis, with characteristic self-confidence, paid no attention and assured Alvarez that as soon as the three services had laid eyes on the Mark I GCA system, they would want working models "yesterday."

Leaving Loomis to fight the bureaucratic battles, Alvarez applied himself to building the Mark I. The system used two trucks parked halfway down the runway, the larger of the vehicles carrying a gasoline-driven generator directly behind the driver's seat. Next came the azimuth antenna, pointing downwind at the approaching aircraft, rocking left and right, per their Ritz-Carlton epiphany. Bringing up the rear was the tall elevation antenna, scanning up and down. Above the generator, he mounted the area search antenna, despite objections from the air force experts that it was unnecessary. Had he heeded their advice, the Mark I would have been a "dismal failure." The GCA truck, parked in front of the antenna truck and facing the landing aircraft, contained the controller's communication system and the screens displaying the radar signals. They completed their first successful trial run in November 1942. After several more weeks of testing at East Boston Airport, where Alvarez worked on improving his "talk down" technique through trial and error, they were ready. On February 10, 1942, General Harold McClelland, director of Air Force Technical Services, requested a formal demonstration of the Mark I to be staged at Washington National Airport.

Four days later, on Valentine's Day, a large group of high-ranking army, navy, and RAF officers assembled to view the test. Unfortunately, the long drive had loosened many of the connections, and Alvarez was forced to postpone twenty-four hours. The next day, the high-voltage tubes in the transmitters kept blowing, and he postponed another day.

The following day did not go any better, and again he had to ask for their forbearance and invited them to return the next day. That night, he and his crew never went to bed, staying up to check and recheck every connection and vacuum tube. As Alvarez greeted the officers the next morning, he was pleased to see that the same group had gamely shown up once again. But just as he was about to begin, Larry Johnston whispered in his ear that the system was down again—more burned-out tubes. The only nearby source of tubes they knew of was Anacostia Naval Air Station, directly across the Potomoc. While Alvarez stalled for time, his pilot took off and flew across the river, returning in only a matter of minutes. With great relief, Alvarez ushered them onto the field, where the military brass listened in via loudspeaker and watched the planes respond:

> We demonstrated that the aircraft were really under my control even though I could only follow them on radar. The high point of the demonstration was the approach of a colonel whom General McClelland had told only to get up into the air, tune his radio to a certain frequency, and do whatever he was told by some voice on the ground. The colonel had never heard of GCA but was an experienced instrument pilot. He first checked me out by changing his altitude and his heading. After each change I told him what he had done. He made several perfect approaches and then landed under the hood.

Just as Loomis had predicted, the three services rushed to order hundreds of GCA sets each. When they heard that ten preproduction units were available, they immediately called a meeting at the Pentagon to determine their equitable distribution in the United States and England. Loomis asked Alvarez to tag along to the meeting, and as he later recalled, although his demonstrations of the Mark I would continue for a week, Loomis' sales campaign "succeeded that first day":

> Neither of us said a word as the admirals, generals, and air marshals engaged in a horse-trading session that ended up with all ten sets allocated to the services, and none to MIT or to the NDRC. The meeting was about to break up when Loomis said quietly, "Gentlemen, there seems to be some misapprehension concerning the ownership of these radar sets; it is my understanding that they belong to NDRC, and I am here to represent that organization." His training as a lawyer

was immediately apparent, and after he had shown in his gentle manner that he held all the cards, an allocation that was satisfactory to all concerned was quickly worked out.

After Alvarez's success with GCA, he went on to invent two other closely related microwave early-warning systems: MEW, one of the Rad Lab's most spectacular inventions; and Eagle, a blind-landing system that had more than its share of trouble getting off the ground but was worth it in the end. Loomis made Alvarez head of his own division, special systems, also known as "Luie's gadgets," and championed his ideas even when many others at the lab had their doubts. Alvarez set to work building a gigantic radar mounted on a circular track that would slowly scan hundreds of miles out over the Pacific to give early warning of the approaching enemy. He finally managed to get his monster antenna to work, and a MEW set with a fifteen-foot-long billboard antenna was operational by mid-1942. An improved set was demonstrated to the Army Air Forces board members in the summer of 1943, and as *Fortune* reported, for the first time they realized what the giant could do:

> They saw a novel array of six scopes on a single radar, with an observer at each. He looked not simply at a few course indications of aircraft, but at clear signals, small blips of light, from each bomber in flight as far as 180 miles on the line of sight from every direction. As each sweep came by, the blips could be seen to move, indicating the flight tracks of the planes. Bunched planes near an airport could easily be resolved into units. Each scope gave accurate, up-to-the-minute data on flights of a huge number of planes. Each operator could concentrate on an assigned wedge-shaped slice of his scope and easily vector (give directional orders to) a plane.

The MEW program was immediately speeded up, and an improved, experimental set was rushed to England, where it was set up at Start Point, on the tall Devonshire cliffs overlooking the British Channel. Rad Lab physicists helped the army assemble the radar set in December 1943, and it was then entrusted to the RAF. The MEW system transformed the inaccurate technique of the old British filter center, allowing all planes to be tracked on the plotting screen, and the pilots linked to the operations room by radio phone. The MEW did not see service until 1944, when its performance exceeded all expectations and allowed the American scientists a moment of pride in the remarkable technological service they had rendered to the British in return for the cavity magnetron.

OVER the course of 1942, the Rad Lab physicists also pushed ahead with Loomis' Loran project. As was the case with most of the Rad Lab's new devices, the navy and air force had little interest in Loran at first and were unimpressed with their initial tests. They blamed the new method for all the errors they found in position finding, even though most of them originated with the old system they used for comparison. In January, a month after Pearl Harbor, another field test was made using Montauk, Long Island, and Fenwick, Delaware, as transmission stations, and Bermuda substituted as a "ship." The Bermuda test turned out to be decisive, establishing once and for all the reliability of sky waves. The average of all the readings agreed with the calculated figure within a microsecond, with an average error of only plus or minus 2.8 miles. They decided that the medium of frequency of about 2.9 megahertz worked best and headed back to the Rad Lab to develop new, higher-powered transmitters, along with improved and simplified transmitter timers. At this point, Loomis was finally able to persuade the United States Navy and the Royal Canadian Navy of the importance of Loran; as one of the physicists on the project recalled, "The submarine menace made it easier to persuade the two navies, particularly because the convoy route from Cape Sable to Ireland had some of the worst weather in the world."

One minor hiccup that occurred in the Loran tests that spring was when the frequency the group chose, 2.2 megahertz, turned out to be the same channel used by a local ship telephone link, and the Rad Lab's powerful transmitter caused phones to ring off the hook all over the Great Lakes area. The navy was not amused and promptly instructed the Rad Lab to abandon the channel. After learning that the frequency of 1.95 megahertz was available—it had been used by ham radio operators before Pearl Harbor—the Loran group quickly claimed it for their own. On June 13, Loomis and the Loran crew conducted the first full-scale operational trial, using a navy blimp equipped with a Loran receiver indicator which was launched at Lakehurst, New Jersey. The trials were so successful in demonstrating the potential of a highly accurate navigation system that suddenly Loran was in great demand: the navy needed it right away for antisubmarine work; and the air force wanted it to help in the ferrying of aircraft across the Atlantic from Brazil to Africa.

A high-level meeting in the Joint Chiefs of Staff Building in Washington was quickly arranged among representatives of the army, navy, and OSRD to discuss the most effective way to apply Loomis' new navigation system to the war effort. Only a few Loran sets had been built by the Rad Lab for research purposes, and there were not enough to go around. It would take several months to produce more, so it was necessary to sort out who would have to wait. "The argument became warm," recalled Bush, who was presiding, "and the officers ignored the chair and went after one another directly. So I tapped the table and said, 'Gentlemen, you seem to overlook the point as you argue; I "own" these sets.' The discussion then became more orderly, and an agreement was reached." It was yet another instance when the military was forced, by the president's mandate under the OSRD, to cooperate with the scientists. Despite the military's reluctance, Bush noted, the final agreement they hammered out over the Loran system "moved us a long way toward mutual respect, out of which only can arise genuine concert of effort in a common cause."

At the navy's request, the Rad Lab began work on the first Loran network, a chain of four stations that would cover the whole North Atlantic from Greenland to Nova Scotia. Over the summer, the physicists rushed to complete and assemble the equipment for the field stations, while a flying survey party of laboratory and navy personnel selected sites in Newfoundland, Labrador, and Greenland. By September 1942, the two Canadian stations were operational, and the southernmost one was synchronized with the Montauk Point station and with its northern mate at the other end of Nova Scotia. One month later, regular Loran navigational fixes became possible, with the Rad Lab scientists, U.S. Coast Guard, and Royal Canadian Navy supervising the sixteen-hour-daily operation of the service. Bad weather and shipping delays hampered the setup of the northern three stations in Newfoundland, Greenland, and Labrador, and during that critical winter the physicists and engineers braved foul weather and U-boat-infested waters to work on the installations. By spring, the entire seven-station system was operational. The navy's next priority was the northeast Pacific, where ships needed help navigating the fog-bound Aleutians, and the northeast Atlantic. Eventually, the Loran network covered the whole of Europe east to the Danube and south to the North African coast.

One of the shortcomings of Loomis' Loran system was its relatively short range over land: 150 to 200 nautical miles for the ground wave, as opposed to 700 to 800 nautical miles over water. Once it was discovered that after sunset sky waves traveled equally well over land and water, a new form of Loran known as SS ("Sky-wave synchronized") was developed. SS Loran appeared to be particularly well suited to nighttime operations, such as those by RAF Bomber Command, which flew planes over central Europe at night, so the navy requested that the Rad Lab start a full-scale trial of the SS Loran system in the United States. Stations were set up near Duluth, Minnesota, and on Cape Cod to establish an east-west baseline; Key West and Montauk were used for a north-south baseline. By fall, army, navy, and RAF observation planes were flying missions across the east-central United States, navigating entirely by SS Loran. The results were "marvelous and phenomenal," according to the pilots who flew the B-18s, and the navy immediately diverted some of its badly needed Loran ground station equipment to the European theater to help the RAF bombers.

Loomis reported on Loran's progress to the secretary of war over a long dinner at Woodley on May 7, 1943. Afterward, the two men sat on the porch, talking late into the night. Stimson noted that his cousin was in "cheerful spirits" and full of news of "the enormous work being done by the laboratory in Boston." Loomis described some of the new inventions and assured him that "the Germans have not progressed nearly as far in their developments of Radar as we have." Their intelligence reports indicated that very few German submarines were even equipped with ASV radar yet. But he again warned Stimson, "Everything depends on our pushing ahead as rapidly as we can before they have developed it."

Riding high on the Rad Lab's string of successes, Loomis agreed to meet with the Rad Lab's official historian, Henry Guerlac, who had been appointed by Bush to write a detailed account of the radar project, intended to justify, in the event of a congressional investigation, the huge sums of money spent. Loomis regarded the whole undertaking with suspicion and saw it as a kind of bureaucratic apologia aimed at mollifying the politicians on Capitol Hill and proving some half-baked thesis about government programs that he personally wanted no part of—"as for example, that capitalism was a bad thing." He made this clear with almost his first remark, Guerlac recalled, when he "brushed aside" his carefully prepared list of questions, being much more interested in interrogating his interrogator: "He is a man of abundant energy, who talks rapidly and confidently, and who dominated the conversation from start to finish so effectively that I never really succeeded in making an interview of it."

Finally, "while trying to avoid all direct references to himself," Loomis reluctantly came round to answering a few questions about the developments in microwave radar "in which he took such a leading part." He was thrilled by the United States' ability to outdistance Germany in wartime radar capability and said that it was "convincing proof of the magic efficiency of American individualism and laissez-faire." He believed Bush's confidence in the ability of civilian scientists to apply their talents to military invention—"leaving them with complete freedom" on technical matters—epitomized the American way. The country's fast results, he argued, came from "free agency and free[dom] from politics." That said, he made some excuse and hurried away, leaving the stunned historian with almost nothing to show for his hour's time with the microwave committee's illustrious chairman.

When Guerlac later mentioned Loomis' dismissive manner to John Trump, one of the other Rad Lab physicists, he was assured that it was probably nothing personal and that Loomis was just making sure he did not "gum up the present work at hand (e.g. building radar to win the war) by writing anything that would offend anybody." Loomis had "one important characteristic," Trump noted. "His ability to concentrate completely on a chief objective," even at the cost of neglecting "matters that appear to other people to be of equal importance."

THROUGHOUT the winter of 1943, the British and American air forces continued carrying out extensive bombing missions over Germany. Now that they had gained control over the Atlantic, and ASV radar together with a strengthened convoy system had broken the back of the German submarine offensive, the Allies could concentrate on the war in Europe. For months now, Allied troops had tried to smash their way into Italy and had been repulsed, and it became obvious that a new plan was needed to unlock the stalemate at the front. Beginning in March, American planes equipped with the H2X blind-bombing radar system began destroying Italian roads and railways that supplied the enemy stronghold at Anzio. Loomis' Rad Lab had delivered the first H2X systems to the U.S. Army Air Forces that fall, and now they would have the satisfaction of seeing it fly blind-bombing missions over Germany. Rabi, who asked of every invention, "How many Germans will it kill?" and had pushed the microwave radar sets from ten centimeters to three, was responsible for making the advanced systems that much deadlier: now bombardiers flying above thick cloud cover could still see, on their radar screens, detailed images of strategic targets on the ground.

The big push began in the summer of 1943, when the U.S. Seventh Army and British Eighth Army invaded Sicily in one of the bitterest and costliest campaigns of the war. On September 3, Italy surrendered, and the Allied armies drove forward, slowly fighting their way north along the peninsula toward the Gustav line, where the German defense held. To break the locked front, the British planned a diversionary attack to give American amphibious forces a chance to make a surprise landing at the beach at Anzio, close to Rome. On January 22, 1944, when the American divisions waded ashore at Anzio, they brought with them the Rad Lab's SCR-584 gun-laying radar systems, which the troops buried deep in the sand so that only the antenna was visible. Although they had trained more on textbook than on actual sets, with the radar guiding their artillery, they shot down five German planes the first night, and before the month was over the total had risen to sixty-three. While they achieved the beachhead at Anzio, the operation—in exposing a large force to risk for a relatively small advantage—was not considered an Allied triumph. But the performance of the Rad Lab's radar in securing the perimeter had been impeccable.

In the early months of 1944, the bombing raids over Germany intensified, preparing the way for Operation Overlord, the Allied invasion of the European heartland. The ultimate success of the Normandy invasion depended on minimizing the strength of the German opposition. The Ninth Air Force had as their objective the destruction of German fighter strength, and flew radar-guided missions over all of northern France, Belgium, and Holland, identifying and attacking airfields and landing strips. Allied bombers also wreaked havoc with the French transport system, taking out the railways, roads, and bridges the German army would need to build up forces at the battlefield. It was expected that D-Day casualties would be high—very high—and the devastation from the steady strategic bombing was the best hope the seaborne infantry had that they would survive the landing and initial combat. In January, Loomis and Stimson discussed the secret operation, and the secretary of war noted that Alfred was full of warnings about rockets: "He thinks they are going to take the place of artillery and, as he is a pretty shrewd in his outlook, I am giving considerable weight to that now, thinking up the possibility of getting a rocket coverage for Overlord."

Under the stormy skies of D-Day itself, the Rad Lab's state-of-the-art radar systems stood watch, guaranteeing the Allied troops fire support during the landing and security from surveillance. On June 6, 1944, the largest amphibious invasion force ever mounted hit the beaches at Normandy. They were accompanied by a total of thirty-nine SCR-584 radar sets, which would help protect the infantry from air attacks as they advanced through Europe. In the darkness of the early morning hours, 450 airplanes equipped with H2X radar systems bombarded the French coastline, cloaking the beach in clouds of smoke and dust as five Allied divisions—two American, two British, and one Canadian—struggled ashore through the surf and dodged enemy fire as they headed for the shelter of the cliffs. It was a precisely timed operation, allowing only five minutes between when the last bomb fell and the first swarm of infantry debarked. While no Allied troops were felled by misdirected bombs, the fear of releasing payloads on their own men compounded a variety of other errors, resulting in hundreds of bombs being dropped onto fields behind German front lines and leaving thousands of American soldiers to be slaughtered at the water's edge on Omaha beach. The air bombardment was more successful at Utah beach, where radar beacons successfully guided parachute troops and glider-borne infantry to their targeted drop zones. The Rad Lab's precise navigation system, known as landing craft control (LCC), was also used to control the landing of the invasion force, directing wave after wave of assault troops to prearranged points on the sixty-mile-long beach.

By evening, a beachhead had been established. Despite the horrific losses at Omaha beach—where the American army suffered most of their 4,649 casualties—the Allies had succeeded in landing 120,000 men with artillery and supplies on the French shores. The Atlantic Wall—miles of trenches, reinforced concrete, barbed wire, and mines blocking access to Germany—had finally been breached. From then on, it was only a matter of time before Allied victory was assured.

The MEW system, set up across the Channel, also had a chance to star on D-Day. Beginning in the prelude, as fighter sweeps were sent over France, the MEW radar tracked their progress and spotted the enemy interceptors that soon followed. The pilots were warned, and as a result, the fighters made unprecedented numbers of kills. It also helped guide bombers over specific targets and aided in the air-sea rescue of downed pilots. Another set, mounted on a truck, plowed through windshield-high water onto the Normandy beach and ran much of the Ninth Tactical Air Command's missions in support of the First Army. When Patton went on the offensive, the British borrowed a MEW set from the lab so the Nineteenth TAC could have it to support the Third Army. Rad Lab scientists had a front row seat on the aerial assault as they worked in the control room, standing by with little or no sleep for days, checking the system, observing the results, and correcting tactics. Alvarez's MEW system, the greatest of the high-power warning radars, ended up chaperoning more and more tactical aircraft—controlling Thunderbolts flying off the Brest peninsula, dispatching bombers, and arranging rendezvous with friendly tank columns. According to *Fortune:* "Many a fighter pilot, returning from the Continent exhausted and out of ammunition, knew he owed his life to the radio voice of the seeing radar eye in the Fairlight ops room, which called out to him some such warnings as 'Bandits on your left, take vector 090.' The longhairs with their giant folly with short waves had brought him home."

On June 12, six days after the Normandy invasion, the first buzz bombs came over the English Channel, loudly announcing themselves before suddenly, silently, plunging to earth, followed by a deafening explosion. In the first month of the V-1 attack, thousands of civilians in London and neighboring cities were killed by the flying bombs, which came to be known as Hitler's "revenge weapon." RAF fighters, guided by MEW, teamed up to fight the V-1 menace and succeeded in destroying a great many of the flying bombs before they hit their targets. Before long, RAF patrols cruised the skies day and night, waiting to intercept the deadly drones, which the MEW system could spot as far as 130 miles away. But the ballistic missile program was the pride of the Third Reich, and they had a seemingly endless store of these rockets, which were capable of placing a few tons of explosives on London daily.

After a desperate plea from Churchill to Roosevelt, two hundred of the Rad Lab's SCR-584 gun-laying radar systems were transferred and deployed against the V-1s. What made the SCR-584 so effective in the end was the miniaturized radar proximity fuse—a shell with a radio-controlled detonator—which exploded near a plane or flying bomb for maximum destructive effect. Along with radar, the proximity fuse was one of the most important applications of the cavity magnetron, and it was given top priority by Bush's NDRC, who assigned the development of the "smallest radar" to the noted Carnegie physicist Merle Tuve. For this revolutionary new device, the flight characteristics of the German V-1, which traveled in a straight line at a constant speed, made for a relatively easy target. Unlike the RAF fighters, which were not always successful in destroying the missiles as they left their launch sites, the SCR-584s, coupled with the proximity fuse, inflicted a heavy toll on the V-1s, destroying 85 percent of the flying bombs that succeeded in crossing the Channel.

On August 12, General Sir Frederick A. Pile, commander of the British antiaircraft command, sent his congratulations to General George Marshall: "The curve is going up at a nice pace, and already we are far away ahead of the fighters. As the troops get more expert with the equipment I have no doubt that very few bombs will reach London." Marshall forwarded the note to Bush, who passed on his thanks to Loomis.

By September 1944, the SS Loran system was up and in regular service over the Continent, enabling nighttime navigation over land and sea. In the final months of the war, it helped with the bombing operations over Germany. RAF pathfinders equipped with SS Loran flew roughly twenty-two thousand bombing sorties. In the preceding months, the pilots had been relying on the Rad Lab's radar blind-bombing system, the H2X, but when the results with Loomis' navigation system were found to be better, the decision was made to conduct all the area bombing operations entirely by SS Loran. Night after night, SS Loran–guided pilots flew strategic bombing missions over the heartland of Germany, raining destruction down on its cities, factories, and railroads. By VJ-Day, the Loran chain extended over one-third of the globe and over most of the contested area, including the Pacific theater, where Loran stations had provided crucial navigational guidance for the Twentieth Air Force's bombing of Japan. Loran transmitters installed in the Himalayas guided traffic over "the Hump" and safeguarded the vital supply routes into China, Burma, and India. Loran would continue to develop as a vital navigation system in peacetime, offering endless possibilities. The total cost of the Loran project from December 1940 to its close was approximately $1.5 million, while an estimated $100 million was spent on Loran systems, including shipping, assembly, and installation. The research and development of Loran came to no more than 2 percent of the government's investment in the equipment—not a bad record for Loomis' laboratory, which under the conditions of war research was "always ready to sacrifice money to buy time."

IN a sense, the Rad Lab was a catalyst for the burst of creativity and inventive effort that would propel American scientists toward their pioneering achievement in Los Alamos. In the early days of the war, it was Loomis, in his role as scientific agitator, who had been the primary force in organizing the country's nuclear physicists to work on radar, at a time when the atom splitters had little to do and fission's useful applications still seemed remote. So by the fall of 1942, when Bush, Conant, and General Leslie Groves took steps to form the highly secret atomic bomb development program, which was then known as the Manhattan Engineering District (later as the Manhattan Project), along with an urgent effort to develop the component elements in sufficient quantity, they had to look no further than Loomis' Rad Lab for a readily available pool of brilliant minds to draw on. These scientists had been collaborating on war research for over a year and would bring with them that cluster of collegiality, friendship, and trust that would help mitigate the terrible pressure and frictions inherent in the task ahead. Moreover, the Rad Lab had grown sufficiently in size and number so that those who were taken away would scarcely be missed, and their projects could be completed on schedule by others who would follow in their footsteps.

As soon as General Groves appointed Oppenheimer as scientific director of the Manhattan Project, he suggested he recruit his top men from Loomis' brain trust of physicists. There would be one critical difference: Unlike Loomis' civilian operation, the Manhattan Project was to be a military lab, which meant inducting the physicists into the army, something many of them were not happy about. Rabi objected vehemently and pointed to their recent experience at the Rad Lab. "We *knew* the military," he explained. "We'd been engaged in making military things, had the military around us. We knew it wouldn't work. In the first place, none of us would come."

After a number of heated talks, a compromise was struck allowing the early experimental phase of the Manhattan Project to proceed under civilian administration along the lines of the Rad Lab, with the military assuming control after January 1, 1944 (the latter actually never occurred). "The first idea that Conant and Groves had was that the bomb was such a hot secret that they should get the boys out there in the fall of 1943," recalled Kenneth Bainbridge, who was among the first Rad Lab alumni to leave for the New Mexico laboratory. "On January 1, they would have to decide whether they would go back and keep their mouths closed forever, or they'd stay on for the duration under the military procedure and put on uniforms."

Once that hurdle was crossed, Oppenheimer was eager to start recruiting his staff and went to Cambridge to begin his raid on the Rad Lab. He knew he would have to coax many of the physicists into leaving Loomis' laboratory, and this might inevitably cause some turbulence. As he wrote Conant: "In view of . . . the very large number of men of the first rank who are now working on that project, I am inclined not to take too seriously the no's with which we shall be greeted. . . ." Oppenheimer began by courting Alvarez, a former Berkeley colleague and old friend. "Great salesman that he was," recalled Alvarez, Oppenheimer had no difficulty convincing him to head west and join the atomic bomb project. "He wanted me back on his team and hinted enough about the challenges of the separation project to persuade me to leave radar." He also set his sights on Rabi, another old friend. But Rabi refused Oppenheimer's overtures and insisted on staying at the Rad Lab to finish his work. In the end, he agreed to serve as a consultant and would be one of the few scientists permitted to go back and forth to "the Hill," as they dubbed the Los Alamos lab.

Loomis was shocked when he first heard Oppenheimer wanted to steal several of his ablest division heads, and he and DuBridge "hit the roof." He had known Oppenheimer since his first visits to Berkeley, and while he acknowledged his brilliance, he had always found him a bit too arrogant and cocky. But out of loyalty to Bush and Lawrence, Loomis finally relented. Over the next few months, the list of scientists who joined the migration from the MIT Rad Lab to Los Alamos grew to include Norman Ramsey, Bob Bacher, Hans Bethe, and George Kistiakowsky. Oppenheimer needed everyone there by mid-April for a conference on the many problems of physics and technique that loomed large ahead of them. One by one, the physicists slipped quietly out of Cambridge, heading out to the desert by train and traveling under assumed names.

By that spring, the details of the uranium 235 bomb, called the "Thin Man," had been pretty well worked out, and their main focus was directed toward the development of the plutonium 239 bomb. Oppenheimer asked Kistiakowsky, who had become an explosives expert at the request of the NDRC, to head up the effort, and he appointed Alvarez as his right-hand man. Fifteen months later, just after dawn on July 16, 1945, the first atom bomb test, Trinity, took place at a barren stretch of desert near Alamogordo, New Mexico. Lawrence, Bush, and Conant had come to the mesa for the demonstration and lay sprawled in shallow trenches twenty miles northwest of the tower-supported bomb. Alvarez was approximately twenty-four thousand feet above ground zero, watching from the cockpit of a B-29 bomber, when the sky suddenly turned bright, and he saw "an intense orange-red glow through the clouds." He had volunteered to measure the explosive energy of the bombs that were to be dropped on Japan, and this was the only dress rehearsal.

As Kistiakowsky watched the ascending mushroom cloud through his welder's mask, he felt the same combination of surprise and relief shared by so many of his colleagues at that moment—the bomb had actually worked. The detonators he and Alvarez had designed had fired as planned. It had gone off without a hitch. He looked around for Oppenheimer to collect his money. In the nerve-racking days before the bomb's debut, they had started a betting pool to help ease the tension, each of them wagering on their estimate of the explosive yield. "Oppie, you owe me ten dollars," he told Los Alamos' director, who had entered a cautiously lowball guess. Later, they calculated the nuclear blast was equivalent to eighteen thousand tons of TNT, a yield of 18.6 kilotons, far greater than anyone had expected. Rabi, who had arrived late and had to take the last bet, which was eighteen—the theoretical maximum—wound up winning the pool.

After their initial jubilation at the outcome of the experiment, Rabi recalled the reverberating wave of dread that followed the "opening of the atomic age":

> While this tremendous ball of flame was there before us, and we watched it, and it rolled along, it became in time diffused with the clouds. . . . Then, there was a chill, which was not the morning cold; it was a chill that came to one when one thought, as for instance I did of my wooden house in Cambridge, and my laboratory in New York,

and of the millions of people living around there, and this power of nature which we had first understood it to be—well, there it was.

The physicists knew the brutal island war in the Pacific made the use of the bomb inevitable. They had wondered briefly if the bomb project would go forward after Roosevelt's stunning death from a massive cerebral hemorrhage on April 14, 1945. It had struck him down in the midst of posing for his presidential portrait, and he had died a few hours later at three thirty-five P.M. Later that same day, Stimson began briefing Harry Truman, the newly sworn in commander in chief. "Stimson told me," Truman wrote in his memoirs, "about an immense project that was under way—a project looking to the development of a new explosive of almost unbelievable destructive power." With Germany's collapse a month later, on May 7, it appeared certain the bomb would be used against Japan. The weapon they had raced to develop to save the free world would be used to destroy a ruthless enemy and terminate the war. The success of the Trinity shot all but guaranteed that the new president would continue what his predecessor had begun. After weeks of debate over how the bomb should be used to bring surrender—a technical demonstration coupled with a warning was rejected by Lawrence, Oppenheimer, Compton, and Fermi as unlikely "to bring an end to the war"—the target cities in Japan were selected.

A few days after the test, Alvarez assembled spares of all the measurement system components, fitted out a toolbox to service the equipment, and made out a will. He packed the uniform and official papers that identified him as an air force colonel, so that if their plane was downed over enemy territory, he would be treated as a military officer and not executed as a spy. On July 20, he flew to Wendover, Utah, and boarded one of the Green Hornet Squadron's transports, which would ferry him to the B-29 base on Tinian island, fifteen hundred miles from Japan. From there he would fly the mission in an escort plane three hundred feet behind Colonel Paul Tibbets' *Enola Gay*, which would carry the Thin Man's smaller and lighter brother, a four-ton atomic bomb known as "Little Boy."

In the end, the calutron that Lawrence sketched in Loomis' living room would supply virtually all of the U-235 uranium used in the bomb the *Enola Gay* dropped on Hiroshima on August 6, 1945. The sixty-inch cyclotron, using a method of gaseous diffusion that was developed at a second wartime plant in Hanford, Washington, would produce the fissionable plutonium 239 for the second bomb, dubbed the "Fat Man" in honor of Churchill, dropped on Nagasaki three days later.

B Y the summer of 1944, Loomis had already begun to look ahead to the end of the war. The Rad Lab, whose experimental microwave technology had once been labeled by the army as "something for the next war," had produced over a hundred distinct radar systems, most of which were in service and helping to speed the day of victory. But for much of the last year, the excitement of developing new radar equipment had been replaced by the laborious administrative task of seeing it produced and mobilized. Loomis spent much of the year chained to a desk in Washington and expediting patent filings so that the radar projects could be moved forward as quickly as possible. It was a tall order, as DuBridge described in a summary report: "By June 1943 nearly 6000 radar sets of Radiation Laboratory design had been delivered to the Army and Navy, 22,000 were on order, and production was climbing past the rate of 2000 sets per month of all types. The total value of Service orders had by that time grown to three quarters of a billion dollars. Production mounted rapidly during the latter half of the year, and equipment with trained personnel were reaching the theaters in large quantities."

Loomis presided over a laboratory that had ballooned into an organization of nearly four thousand—five hundred of them physicists—with emissaries all over the world implementing its war-winning ideas and devices. It had sprawled over more than fifteen acres of floor space in Cambridge and elsewhere, spent approximately $80 million in federal funds, and in its last year had reached a budget of about $125,000 a day, or close to $4 million a month. It was, in the words of Karl Compton, "the greatest cooperative research establishment in the history of the world." While many found fault with the administrative eccentricities of the organization—or "dis-organization," as some critics maintained—it was General Patton, witnessing the Rad Lab's radar systems in operation in the Rhineland in 1944, who observed, "This is the way that wars not only can, but must be, run from now on." Loomis had helped to force the development of radar within the army, and in the opinion of many of his peers, his greatest contribution lay in his brilliantly orchestrated effort with Stimson to mobilize the products of science and technology, break down military resistance to the flow of innovative ideas and applications, and continuously press for further experimentation and the acceptance of new weapons systems and tactics. As Lawrence told an interviewer at the time, "If Alfred Loomis had not existed, radar development would have been retarded greatly, at an enormous cost in American lives."

Lawrence, who had stood at Loomis' elbow in the early days of the laboratory, could not pay high enough tribute to the banker-turned-scientist who had organized physics for war and exerted his enlightened influence on the kind of war the country was ultimately able to fight: "He had the vision and courage to lead his committee as no other man could have led it. He used his wealth very effectively in the way of entertaining the right people and making things easy to accomplish. His prestige and persuasiveness helped break the patent jams that held up radar development. He exercised his tact and diplomacy to overcome all obstacles. He's that kind of man." Lawrence added, "He steers a mathematically straight course and succeeds in having his own way by force of logic and of being right."

As visionaries of the wartime laboratory, Loomis, Lawrence, and Compton had had the hubris to hire a staff dominated by physicists, and it enabled them to create an environment for research run for, and almost completely by, physicists, with everything subordinated to preserving their freedom and creativity and the production of their revolutionary technological devices. In doing so, Guerlac wrote in his official history of radar, "the Radiation Laboratory came close to realizing a scientist's dream of a scientific republic, whose only limitation was the supply of scientists." There were those who believed that the laboratory's great success story should continue on after the war and that still more marvelous gadgets and techniques might be forthcoming. Loomis was vehemently opposed to the idea and took decisive steps to stop the juggernaut, paying a call on President Roosevelt. Loomis felt the Rad Lab would surely stagnate and falter and argued that "only the pressure of war" could make a government program of that size and magnitude flourish. He had great faith in private enterprise and a deep suspicion of public ones. While acknowledging that his opinion appeared to conflict with recent experience, Loomis maintained there was no cause short of the national defense that could have inspired him to help create a large, centralized, government-controlled laboratory, and the very idea was "anathema" to him. War was a great stimulus to science, but it was not a stunt that could be repeated in peacetime. Bush shared his views, and it was decided that the Rad Lab should be terminated.

With his war job almost over, Loomis was eager to return to private life. He felt worn down. He was exhausted by the constant travel and still suffered from the lingering effects of a serious bout of pneumonia. His marriage, which for years had existed only in appearance, was now at an end. His affair with Manette Hobart, long confined to the Glass House and furtive meetings at hotels, had become an open secret. There was no longer any question of his returning to his old home on Club House Road in Tuxedo Park. Throughout the war years, he had repeatedly packed his sickly wife off to sanitariums for her health, and Loomis, in a rare miscalculation, made the mistake of trying to have her committed permanently. When his oldest son, Lee, returned from war and discovered what his father had done, he rushed to his mother's defense.

Lee was a big, obstreperous young man, and although he had followed his father's lead and graduated from Harvard Law School and eyed a Wall Street career, the two had never gotten on. They had always knocked heads, now more than ever. Lee became his mother's self-appointed protector and guardian. He took control of his mother's half of the Loomis fortune and shrewdly invested it on her behalf. Both Lee and Henry regarded what their father had done as traitorous and angrily broke off all relations with him. The middle brother, Farney, refused to take sides. While Henry eventually reached an uneasy truce with his father, Lee would have scarcely anything to do with him for the next twenty years. "It was a very bitter divide," said Lee's daughter, Sabra Loomis. "They didn't speak."

Like most members of the family, she only knows bits and pieces of the story, because no one ever talked about it. During the war years, she lived with her grandmother in Tuxedo Park off and on when her parents were away, and they were very close. By then, Ellen had taken up the use of her maiden name again and signed her letters Ellen Farnsworth Loomis. "I had a dim sense as a child that a wrong had been done to her," recalled Sabra. "I know at one point she'd been shut up in a hospital, and no one was allowed to write to her or talk to her. There seemed to be a bit of collusion going on in that the doctors had said, 'No visits from the children, no calls from the husband, no visitors,' and she thought she had been dumped there and abandoned. Alfred was very powerful and could have what he wanted. I always thought that there must have been something that precipitated it, whether it was that Alfred had already deserted her and she found out, but I don't know. Only that if she was depressed when she went in, she was much more depressed by the time she got out."

For someone who had always prided himself on being plainspoken, Loomis had been less than candid with his three sons about the existence of another woman. When he first broached the subject of the divorce with Lee, he denied there was anyone else. After they learned the truth, his deception made it seem that much more terrible. "They were all in a state because Alfred had lied to them," recalled Paulie Loomis, who was engaged to be married to Henry. "Divorce was a terrible thing in those days, and between that and having Ellen locked up, he halfway destroyed her. Ellen blamed herself, and after that she just started to fall apart. I think Alfred did it just to get her out of the way," she said, adding, "It's the only thing I ever held against him."

In the fall of 1944, Manette and her two young boys moved out west and took up residency in Nevada, as the state regulations governing divorce at the time required. Hobart did not contest the divorce, and in February she signed the papers making it official. Manette remained out west for the rest of that winter while she waited for Loomis to extricate himself from his marriage. He stayed at a neighboring ranch and commuted back and forth to Washington and Cambridge. On April 4, 1945, Alfred and Ellen Loomis' divorce was final. A few hours later on the same day, he and Manette were married. A justice of the peace in Carson City, Nevada, performed the brief ceremony. No friends stood up for them, but a photograph taken after they exchanged vows shows Loomis, dressed in a banker's pinstripes, standing stiffly beside his new bride. Manette is smiling shyly up at him and is wearing a prim black suit with a white rabbit-fur collar, cuffs, and matching fur muff, which she had purchased at Bonwit's expressly for her wedding day. He was fifty-seven; she was thirty-six.

Although Loomis provided very generously for his former wife, giving her more than half of his substantial fortune as well as the house in Tuxedo Park and the penthouse off Fifth Avenue, New York society was appalled. It was hard to know what was more unforgivable—the tandem divorces, the hasty remarriage to a much younger woman, or the suggestion of a long clandestine relationship with his best friend's wife. Cholly Knickerbocker, the reigning gossip of the day, wrote a scathing account of the affair in his regular column, "The Smart Set":

It's amazing how many members of the Rarified Set, whose very con-
volution in their social circles generally is noted for posterity by the
scribblers, still can manage to draw the blinds on their glass houses
and keep in the dark the changes that go on within. For example, I'm
sure many of my eager readers are serenely unaware that the senior
Alfred L. Loomis' marital rapture of well over a quarter of a century is
now a very definite rupture—and what's more has been conclusively
phfft for seven months. . . .

The story went on to note that the first public tip-off that the prominent couple were no longer "pulling together in double harness" was the
publication of their separate entries in the most recent edition of *The
New York Social Register*. "What's more, it blandly noted the fact that
Alfred is blissfully enjoying a second Darby and Joan existence with
Belgian Manette Seeldrayers Hobart. . . . Evidently Al long had been a
close friend to Manette and Garret Hobart—for a little research reveals
the fact that when their first [actually, it is their second] son was born in
1937, Manette named the child Alfred Loomis Hobart!"

It would be impossible to overstate the reverberations Loomis' divorce had in the elite financial and social circles he had once frequented. "Oh, I think it was the most shocking divorce at that time,"
recalled Lynn Chase, whose husband, Edward L. Chase, was close to
George Roberts, the head of Loomis' old law firm, Winthrop & Stimson—which, in a move that was regarded as the coup de grâce in the
scandalous affair, chose to represent Ellen and not Alfred in the divorce.
"People absolutely turned their backs on him and had nothing to do
with him for years. It was a combination of the fact that everyone had
been so devoted to his wife, and that she was unwell and could not cope
on her own. Then he had married somebody that had more or less been
in his employ, had worked as some sort of secretary at his laboratory—it
was all very, very shocking. He was like a nonperson after that. He just
disappeared from society."

The newlyweds steered clear of New York for several months, honey-
mooning at Del Monte Lodge, and in Carmel and San Francisco. Ernest
and Molly Lawrence hosted a Saturday afternoon cocktail party in their
honor and invited Don Cooksey and all the cyclotroneers. They celebrated later that night over dinner at Trader Vic's, their old haunt, and
organized a big picnic lunch at Muir Woods on Sunday. While they all
wanted to be happy for Loomis, for whom they had tremendous admira-

tion and affection, the Berkeley scientists, many of whom had met Manette on visits to Tuxedo Park when she was still Mrs. Garret Hobart, were every bit as astonished by the turn of the events as Loomis' old club crowd. It is clear from the correspondence between Loomis and Lawrence throughout this period that even his closest colleague and friend failed to detect that anything was seriously amiss in his marriage. Lawrence had seen Ellen last in the fall of 1944 and had written thanking them for their hospitality—"it was certainly a real treat visiting you and Ellen again"—and expressing his delight that they were planning on "coming out in the spring."

One can only imagine his surprise, to put it mildly, upon learning that a new Mrs. Loomis would be accompanying him on that trip. But the war had disrupted all their lives, and the years of all-consuming research, exacerbated by the burden of distance and secrecy, had taken a heavy toll on many marriages. If not exactly approving—Lawrence's wife, Molly, and Ellen had become quite close—they were not inclined to judge him, either, and welcomed his new wife with open arms. Manette may have discerned a certain reserve on the part of some of the wives and once remarked that she did not get to know Molly very well in those first few years because she was "so wrapped up in her children that she didn't have time," though whether that was by necessity or by design is unclear.

In any case, they usually saw Lawrence alone. He and Manette got along famously and quickly formed a very close bond. She encouraged him to take up painting to relax and later made him pose for a large bronze bust she did of his handsome head. "We always used to joke that I was his teacher and he was my pupil," recalled Manette, who regarded him as far more "human" than most scientists. "We could tease him; he loved that. He loved having a good time. He was full of life. He loved going out with people, and dancing with pretty girls when they went along." Lawrence continued to visit Loomis in New York whenever he could and frequently accompanied them on jaunts to Jamaica and Balboa. The three of them reveled in one another's company, and Lawrence and Manette would flirt outrageously with each other, much to Loomis' evident pride and amusement.

On their return to New York, the couple moved into a large apartment in the Mayfair House, a sort of residential hotel that provided all the amenities, including room service and housekeeping, which Loomis, who had spent the war years in hotels, had come to appreciate.

Manette's domestic skills were also minimal, so it suited them both perfectly. Loomis also purchased a lovely summer home in East Hampton, Long Island, not far from R. W. Wood's old farmhouse, with plenty of room for Manette's sons when they were home from boarding school. But nothing prepared them for the inhospitable climate they had returned to. Most of the extended Loomis and Stimson tribe were not on speaking terms with him, and few of his blue-blooded peers wanted anything to do with his foreign wife. They got a decidedly chilly reception at the exclusive Maidstone Club in East Hampton. "I think it cut him very deeply," said Lynn Chase, who can still remember the way people publicly snubbed Manette. "Here he had all this money and power, and he could not buy the approval of the Maidstone community."

LOOMIS spent the next few months overseeing the dismantling of his laboratory and winding up millions of dollars in contracts. Things had been shipped in all directions, and many scientists were so anxious to help that they would send something to the Pacific without making any record of it. Thanks to Gaither and his crack team of administrators, every piece of equipment would be accounted for. "Out of the $30 million in outstanding claims for the government, MIT did not have to pay one cent," Loomis later observed, "because he realized and I realized the government can thank you for doing something in an emergency, but along comes an order that wants to know where these typewriters are. . . ."

For most of the people who had worked there, it was hard to believe that an institution as big and vital as the Rad Lab could simply be shut down, but Loomis did just that. Never one to be sentimental, he scoffed at those who were reluctant to lock the doors and turn off the lights. By the end of summer, the penthouse sheds had been stripped of their antennas and knocked down, and most of the main laboratories had been emptied, with only marks on the linoleum to show where the furniture had been. Building 22, where Alvarez had done some of his finest work, was closed off, and carpenters were already at work turning it back into dormitory space for incoming MIT students. Like most endings, the Rad Lab's was not nearly as glorious as its beginnings, when it opened for business in the fall of 1940, fourteen months before Pearl Harbor, on the strength of Loomis' vision and the three dozen physicists who shared it.

In the end, after "five years of furious technology," the atom bomb stole its thunder. "On the evening of August 5, 1945," the official year-book noted, "the Laboratory found itself in the same position as the overwrought butler":

> It had worked in secret. Newsmen had long since despaired of a story. But now it was to be told. An open invitation went to the press and newsreels. "Guided Tours" were set up for the next day. A painstaking "news release" was written. And a little before the 7:30 PM deadline, a messenger in a special car was sent off to the news offices of Boston. At 7:15 the messenger phoned in. She had delivered half of her releases, and could deliver no more. Her car was blocked by people running around in the street and kissing one another. It appeared that at 7:00 PM Japan's surrender had been announced.

Although the press covered the laboratory later, the thrill was gone, and its glorious achievements got short shrift. *Time* magazine's scheduled cover story on radar was bumped to page seventy-eight, and the new cover, celebrating V-J Day, credited the work of the Los Alamos physicists. The men who worked on the atomic bomb were hailed as heroes, and countless books and Hollywood movies would recount their exploits, while the daring and inventive minds who created radar were largely forgotten. The Manhattan Project became world famous. The Tizard Mission faded into obscurity. Only the Rad Lab veterans knew better, knew that if radar had not kept the Germans from defeating England, the war might have been over before America entered the contest. Everyone who had worked at the laboratory understood the decisive role their deadly devices had played in speeding the day of victory, and it was reflected in a remark by DuBridge that became something of an unofficial slogan, their badge of honor: "Radar won the war; the atom bomb ended it."

In a strange sense, it was exactly the conclusion Loomis would have written himself. For the record, the yearbook tallied their successes: the lab had begun as a gamble, and it had paid off. They had started out behind and finished ahead. They had made history, smashed the U-boat, and shot down German planes and V-1s. Along the way, they had introduced some revolutionary concepts into warfare and significantly advanced knowledge in the field, packing decades of radar development into only a few years. They had also given birth to a new billion-dollar industry, and at least half a dozen companies were either stating or implying in their advertising copy that radar was their own private invention. While proud of everything they had accomplished, Loomis and his physicists were personally "embarrassed by the problem of telling what they had done," an awkwardness that was reflected in the Rad Lab's perfunctory official statements in the days and weeks that followed. Of course, it was impossible to overstate their debt to the British, not to mention prior work done by the U.S. Navy and Army Signal Corps, and therefore difficult to know how much to lay claim to, not to mention the perennial problem of sorting out who did what in the white heat of the moment.*

The Rad Lab formally closed on December 31, 1945. Most of the physicists returned to their university jobs and resumed their careers as professors and research scientists. After completing all his administrative duties as head of the OSRD's radar section in 1947, Loomis returned to his former activities as a philanthropist, withdrawing quietly into private life. Almost from the moment the Rad Lab ceased operating, Loomis began to disappear. He refused requests for interviews and photographs, and proved so elusive that to get his portrait to accompany their glowing account of his adventures in business and science, *Fortune* magazine had to pursue him by plane all the way to his private twenty-two-thousand-acre retreat on Hilton Head Island. He turned down prestigious job offers and university appointments, including a long-standing offer from Lawrence to come work with him at his Berkeley laboratory. After the war, he was besieged by letters asking for his support for various scientific causes and research projects, along with innumerable invitations to speak before civilian groups—"this Rotary club, that Women's auxiliary," as he put it—most of which he ignored. The requests to sign his name to various protests and petitions were promptly tossed in the waste bin.

Except for occasional appearances at various advisory committee meetings, including an Atomic Energy Commission Panel on Radiological Warfare and the Joint Research and Development Board headed by Bush to counsel the army and navy on strategic matters, Loomis was absent from the Washington scene where he had only recently been such

---

*Loomis' application for the Loran patent was disputed by two scientists, each of whom claimed to be the first inventor of the fundamental concept. The U.S. Patent and Trademark Office twice held in Loomis' favor. The court of customs and patent appeals later reversed the ruling in one of the cases. A Loran patent was finally issued to Loomis on April 28, 1959.

a forceful presence. He was, by disposition, an extremely understated man who really did not care for being center stage. A large part of his success as the laboratory's leader had been his charisma and persuasiveness, a positive thrust that enabled him to win the confidence of so many brilliant scientists and convince them that supporting and furthering their work was his only goal. While he had teamed up with Lawrence as a pioneer of "big science," organizing massive industrial and government funding for his large-scale projects, and in the process changing forever the expenditures of money and manpower that would be committed to such research efforts, his true allegiance was always with "little science." He wanted nothing more than to return to the solitary wizardry of men like R. W. Wood, lone experimentalists who, working practically by themselves in a private laboratory, succeeded in making major contributions to the frontiers of knowledge.

Loomis followed his passion for science to Washington, and then into war, but political influence was something that neither interested him nor held any allure. He did not care to join the ranks of physicists-turned-elder statesmen who were trotted out at conventions and government seminars, to be "exhibited as lions at Washington tea parties," as the distinguished physicist Samuel K. Allison described "the awe and gratitude of the scientifically illiterate lay world." Independence was a luxury he could afford, and it enabled him to remain detached, and slightly above, the postwar scramble for position and power that consumed so many of his colleagues.

Although he attempted to avoid attention and public recognition wherever possible, he continued to collect laurels. There was another honorary degree—this one from Wesleyan. In February 1948, while out in California visiting Lawrence, Loomis received a letter from the British embassy informing him that he was to receive one of their country's highest decorations:

> It is with great pleasure that I inform you that the King has been pleased to award His Majesty's Medal for Service in the Cause of Freedom in recognition of the valuable services you rendered to the Allied War effort in the various fields of scientific research and development.

That spring, Harry Truman awarded him the Presidential Medal of Merit, the highest civilian award, for his contribution as one of the leading scientific generals of the war. In the ceremony on Governors Island on the morning of June 23, 1948, Loomis was cited for his "exceptional meritorious conduct" and the "performance of outstanding services" to the United States from June 1940 to December 1945:

> Dr. Loomis, as Chairman of the Microwave Committee of the National Defense Research Committee, early foresaw the military possibilities of microwave frequencies for radar detection and ranging. His personal qualities and enthusiasm enabled him to enlist the services of many brilliant physicists and engineers in the coordinated development of this new art. . . . A brilliant experimentalist endowed with extraordinary foresight, Dr. Loomis was a central figure in this development program that contributed so significantly to the successful termination of the war.

At the close of the ceremonies, General Courtney H. Hodges told the scientists who were being honored that day, among them pioneers of rockets, antiaircraft weapons, and infrared equipment, that he understood their reluctance toward "the wholesale transference of intellectual effort to destruction." He assured them, however, that they had a distinguished precedent in the great mathematician Archimedes, who turned his genius to the defense of the Greek city of Syracuse and destroyed the invading Roman armies with his fireballs, mirrors, and ingenious instruments of violence.

The allusion to Olympic glory was wasted on Loomis, who suffered from no guilt or lingering doubts about his part in developing weapons of war. Unlike many of the scientists who worked on the bomb, he did not regret the "atomic age" or question the morality of devising even more fearful devices. In fact, Loomis felt there ought to be more courage in experimentation in nuclear physics than before and always expressed great faith that scientists could see to it that their products were used responsibly and to the benefit of mankind. He believed in exploring new scientific knowledge to its fullest extent, moving forward without fear of where the experiments might lead. Nobody could foresee all the possibilities and how they might be applied not only to war, but also to peacetime and utilitarian purposes, with incredible potential advantage to civilization. He could not imagine that any "true scientist" could feel differently: "If you want to find the truth, you must continue to experiment." It was the optimistic credo he had believed in thoroughly all his life. He saw no reason to abandon it now because some people trembled at the awesome power of a nuclear explosion.

Loomis was drawn into the debate over the further testing of nuclear weapons as well as other postwar developments, but increasingly from the remove of his East Hampton home. He made himself available for consultation on his personal opinions but did not care to serve as a spokesman for any particular group or ideology. Leaders in government and industry continued to seek out his counsel because of his long record of success and the almost prophetic accuracy of his appraisals of both men and events. He seemed at times to possess a "seerlike vision," observed Caryl Haskins, who would succeed Bush as president of Carnegie. It was, he wrote, "an insight given only to the greatest of men."

For the most part, Loomis disparaged politics and thought it a great pity that so many good scientists were squandering their time and energy on policy problems when they could be pursuing fundamental research instead. He continued to be a close adviser to Lawrence on all matters and cautioned him to avoid getting caught up in wasteful political activity. He even asked Gaither, who had returned to his legal practice in San Francisco, to keep an eye on his generous friend and make sure he did not fall victim to pressure groups. Loomis made a rare exception to this rule a year later when Gaither was asked by the air force to organize the Rand Corporation, a nonprofit outfit that would apply the "best scientific abilities and achievements" to ensure the national defense, and pleaded with Loomis to become founding trustee. He was finally persuaded and was tremendously influential in the pioneering phase of the organization and later even brought Lawrence onto the board.

Loomis never reopened his Tuxedo Park laboratory. At one point, he looked into the possibility of donating the lab to the Rockefeller Foundation or similar nonprofit outfit, but the Tuxedo Park Association was adamantly opposed to the continued operation of a research facility within its confines. He tried to get more than one neighbor to take the property off his hands and found he literally could not give it away. By the end of the war, Tuxedo was in a terrible decline. Old-time resorters had deserted it, and with more than half of the sprawling "cottages" vacant and run-down, it had become known as "the Graveyard of the Aristocracy." As Cleveland Amory observed in 1948, "No other community in this country ever started off on a grander social scale, and therefore no other may be said to have fallen so hard." Loomis finally sold the Tower House to a developer, who renovated the property into separate rental units and renamed it the Villa Apartments. Almost immediately thereafter, the Tuxedo Park by-laws were changed to prevent the conversion of any other historic mansion into condominium complexes. The Tower House's Tudor facade with its single tower remains substantially unchanged, however, and the dark, ornate entrance hall still seems haunted by old ghosts. His beloved Glass House was purchased by a park resident and has been preserved as a private home, its stark white design and double glass walls testifying to the bold ideals of a bygone era.

In his memoirs, Alvarez, who called Loomis "the last of the great amateurs," lamented that his enormous contributions to science and his country would not be remembered, while conceding that it was "an anonymity of which he would have approved but which hardly does him justice." But Loomis had no interest in assuming an elevated mantle. Taffy Bowen, who had often marveled at the mysterious way Loomis, "as if by magic," knew exactly how to get a project started or where to obtain the requisite materials, discovered years later the lengths his American friend had often gone to in order "to avoid taking credit for the developments with which he was associated." So Bowen was not surprised that when it came to his role in the construction of Lawrence's giant cyclotrons and the development of microwave radar, "he took pains to see that his part in it was covered up." Looking back, he wrote, "the extraordinary thing is the modesty with which all of this was done."

"He didn't take credit for things, that was very characteristic of him," said Haskins, who counted himself among the "fortunate band" of scientists privileged to call Loomis a friend. "Of course, he was known in closed circles, but not widely known, after the war. History forgot him. Well, in a sense he forgot himself, because he didn't care about all that. He wasn't interested in the past. He was interested only in the present and the future."

Epilogue

> He's probably the only man who ever on the one hand took the guys down in Wall Street for a ride and made a lot of money out of them; and on the other hand got elected to the National Academy of Sciences on the basis of his accomplishments in physics.
>
> —Vannevar Bush

Alfred Lee Loomis made it clear he had "no interest" in becoming a science adviser to President Truman when William Golden approached him about the job in the fall of 1950. In the wake of the Korean War, Truman was under pressure to reactivate Bush's OSRD, and Golden had been engaged to advise the president about organizing the government's scientific efforts, and with interviewing likely candidates to fill the job. "He was very gentlemanly and cordial, but it was certainly apparent he wasn't interested in engaging in any new activities," recalled Golden. "He had received some negative publicity after his divorce, and I think he simply chose to withdraw from the world. It was as if some vital ingredient had drained out of him." Loomis remained personally involved with the work being done by his former colleagues and generously supported their pet projects at MIT, Berkeley, and Cal Tech, where DuBridge served as president for more than twenty years. After the war, he returned to his early love of astronomy and became actively involved in funding and building observatories, such as Walter Roberts' High Altitude Observatory in Colorado. He never stopped collecting brilliant men. In the late 1940s, he met an engineer named Avery Fisher, who had a small stereo store on 42nd Street in New York. "They became great friends right away, like two little boys," recalled Fisher's wife,

Janet. When Loomis learned Fisher was having trouble getting financing for his new high-fidelity speakers, he took him straight up to the office he kept with Thorne and instructed his partner to set him up with "an open line of credit." It was the beginning of Fisher Electronics and another adventure with a pioneering scientist. Years later, as Fisher prepared to enter the new Avery Fisher Hall at Lincoln Center on opening night, he told his wife, "Alfred should be here." Loomis enjoyed being surrounded by the "old gang" and, as Alvarez put it, "introducing his scientific friends to the pleasures that are normally known only to the very wealthy." Each spring, he invited Lawrence, Alvarez, McMillan, and DuBridge, along with their wives, to be his guests at the Del Monte Lodge in Pebble Beach and to play golf at Cypress Point. He also hosted annual winter getaways to Jamaica and footed the bill for all the expenses and first-class airfare. Loomis died of a stroke at eighty-seven at his East Hampton home on August 11, 1975. His last obsession, noted Alvarez, who visited him shortly before his death, "was programming tricks for the Hewlett-Packard model 65 handheld computer that was his constant companion."

Ernest Lawrence and Loomis continued their close association, and Loomis was a key behind-the-scenes figure in the Berkeley physicist's postwar campaign to assert himself as the leader of the country's nuclear establishment. Both were advocates of the hydrogen bomb and subscribed to the straightforward view that "we should build it before the Russians do." To that end, Lawrence teamed up with Edward Teller, a longtime champion of the H-bomb, to turn the Rad Lab's Livermore, California, site into a second nuclear weapons laboratory and develop more powerful and efficient accelerators, such as the "bevatron." This led him into direct conflict with Robert Oppenheimer, the director of the Los Alamos laboratory, who was philosophically opposed to an escalating nuclear arms race with Russia. At the heart of the struggle was Lawrence's next "big machine"—the material testing accelerator, or MTA—an ill-conceived plan to build a giant accelerator that would turn out vast amounts of plutonium for the country's nuclear arsenal. Oppenheimer came out against it, and the harder Lawrence pushed, according to the historian Nuel Pharr Davis, the more "the MTA permeated and inflamed all the other issues."

In 1953, as a result of the heightening political tensions, Oppenheimer was accused by Senator Joseph McCarthy and the House

Un-American Activities Committee of having Communist sympathies, and his security clearance was revoked. The bitter dispute over Oppenheimer's loyalty divided the scientific community and strained many of the old Rad Lab friendships. Among those supporting Oppenheimer were Bush, DuBridge, Rabi, Bethe, and Conant. On the other side was Lawrence's camp, including Loomis, Alvarez, McMillan, and Teller.

Loomis, who was conservative and fiercely patriotic, advised Lawrence against testifying at Oppenheimer's hearing. Lawrence decided not to and urged Alvarez to do likewise. "[Ernest] didn't think what was the most popular thing to do," Loomis said later. "He wanted to concentrate his time on scientific research and not get caught up in arguments about whether you're a loyal American. . . ." Alvarez testified as a government witness and, while never directly casting doubt on Oppenheimer's loyalty, criticized his decision to veto the H-bomb. In the end, Oppenheimer was ousted from power and publicly disgraced. Lawrence's group was seen as greatly contributing to Oppenheimer's downfall, and many scientists would never forgive them.

Lawrence and Loomis collaborated on various other projects together, and in 1950 Lawrence came up with the idea for a color television tube while staying with Loomis at his Mayfair House apartment in New York. On Loomis' advice, he and Rowan Gaither formed a company to exploit the various television schemes they developed, and Chromatic Television Laboratories was incorporated and patent applications filed. Lawrence's tube was competitive with one designed by RCA at the time, but the problems involved in financing the tube and getting it into production proved more of a challenge than they expected, and they eventually sold the rights to Sony. Lawrence continued to suffer from frequent infections and ulcerative colitis, and Loomis, in an effort to keep his friend healthy, tried to take him away on vacations and arranged for him to have a full-time chauffeur. On August 26, 1958, Lawrence underwent surgery in a Palo Alto hospital. He died the following morning without regaining consciousness. He was fifty-seven. Only days earlier, he had joked with Loomis on the phone about their next trip: "I'll be all right, all I have to do is get down to Balboa."

Vannevar Bush continued as president of the Carnegie Institute until 1955. A year before he retired, Bush helped mount a vigorous defense of Oppenheimer and denounced the hearings for placing a man on trial for expressing strong opinions: "This board has made a mistake . . . a serious one." Although he disagreed with his Berkeley colleagues, Bush never allowed it to interfere with their relationship, and he, Lawrence, Loomis, and Alvarez remained lifelong friends. Bush's star waned after the war, but he served on various boards, and for many years was chairman of the MIT Corporation, where he managed to coax the occasional large check out of Loomis for new gadgets and laboratory equipment. Bush died in June 1974.

Luis Alvarez, whom Loomis called one of his "other sons," was awarded the Nobel Prize in physics in 1968. He won for his development of the hydrogen bubble chamber into a powerful instrument that made possible valuable new discoveries in the field of high-energy physics. Alvarez died in August 1988.

Karl Compton retired from MIT in 1948 and died six years later in June 1954. His brother, Arthur Compton, returned to academic life after the war and served as chancellor of Washington University in St. Louis from 1945 to 1953; he devoted his final years to teaching until his death in March 1962.

James Conant and Oppenheimer fought hard to prevent the building of the hydrogen bomb, and Conant's feelings for Lawrence and Loomis were never the same after the loyalty hearings. As he said in his testimony, if opposition to the H-bomb made Oppenheimer a security risk, "it would apply to me, because I opposed it—as strongly as anybody else." He left Harvard after twenty years as university president—as he wrote Kistiakowsky, "long enough to serve a sentence for youthful indiscretion"—to become high commissioner of Germany. He went on to become a leading educator and wrote widely on the need for better public education and testing "to break down social barriers." Conant died in February 1978 and was cremated and buried in his wife's family plot in Mount Auburn Cemetery in Cambridge, Massachusetts, where Bill Richards was also laid to rest. His son, Ted Conant, took custody of Richards' roman à clef and the confiscated short story about the uranium bomb, which he eventually passed on to his daughter.

Ed McMillan shared the Nobel Prize for chemistry with Glenn Seaborg in 1951 for their discovery of plutonium (with Kennedy and Wahl) as well as his discovery of neptunium with Abelson in 1940. His work led scientists all over the world to create and identify another fifteen transuranic elements. He returned to Berkeley after the war and worked with Lawrence on developing the next generation of cyclotrons. He died in September 1991.

George Kistiakowsky, whose career Loomis had fostered since his days as a Tuxedo Park research fellow with Bill Richards, went on to become one of the most influential scientists in the country. In 1959, he became scientific adviser to President Dwight Eisenhower. He remained extremely close to the Conant/Richards family and used to take the author sailing off Cape Cod in Massachusetts, where he and his wife, Elaine, had a summer home. Kistiakowsky wrote a memoir of his days at the White House during the intricate test ban negotiations with Russia, and he was working on his autobiography at the time of his death in December 1982. He kept a copy of Richards' novel, *Brain Waves and Death*, in a drawer by his bed.

Robert Wood spent fifty years as a professor at Johns Hopkins. Loomis continued to visit his brilliant mentor at his farmhouse in East Hampton until his death in 1955.

Taffy Bowen immigrated to Australia after the war and launched an ambitious radio astronomy program. In 1954, Loomis helped persuade the Carnegie Institution trustees to contribute $250,000 so that Bowen could realize his dream of building a 210-foot radio dish in New South Wales. A year later, Loomis had a hand in convincing the Rockefeller Corporation, where he was also a trustee, to put up another $250,000, with the stipulation that the Australian government provide a matching grant. Bowen's radio telescope was based on the design of the six-foot SCR-584 gun-laying radar that the British and American physicists had built at the MIT Rad Lab during the war. The construction of the Parkes telescope took six years, but when it was completed, it was the largest device of its kind in the Southern Hemisphere. It opened up a whole new branch of astronomical study, tracking celestial objects across the sky with a precision once thought to be unattainable. He died in August 1991.

Garret Hobart III became a recluse after the war and rarely ventured out of Tuxedo Park. "He was a vulnerable person," said Al Hobart. "After

Manette left him, he was virtually a hermit." He died of cancer at age fifty-five in March 1963. He is survived by his two sons, Garret Hobart IV, a retired lawyer, and Alfred Hobart, who runs the Green Mountain Valley School in Vermont, which trains Olympic athletes.

Manette Seeldrayers Loomis and Alfred enjoyed thirty happy years together. Time never diminished his love for her. His granddaughter Jacqueline Quillen, who lived next door in the house Alfred purchased for her family, recalled that when she stopped by one day to borrow some eggs, he told her: "You can have anything that's mine, except Manette." Late in life, Loomis worried about how his young wife would cope after he was gone. He asked a close friend, Ronald Christie, who had been one of his favorite young protégés at Tuxedo Park and was a prominent lung specialist at the McGill University in Canada, "to take care of Manette." A year after Loomis' death, Manette and Christie were married. Manette died in October 1991.

Ellen Farnsworth Loomis and Alfred did not speak for almost twenty years, until the suicide of their seventeen-year-old granddaughter, Ellen "Debbie" Loomis, the daughter of Betty and Farney Loomis, who plunged to her death from a patio of their old penthouse apartment on January 21, 1965. Alfred rushed to his former wife's aid as soon as he heard the news. The teenager had been under psychiatric care for depression. A note was found pinned to her coat. The funeral helped heal the rift that had kept the family apart for so many years. Ellen died at age eighty-six in December 1975. Her ashes are buried in the small graveyard behind St. Mary's Episcopal Church in Tuxedo Park.

(Alfred) Lee Loomis Jr. became a successful venture capitalist and was involved with domestic oil and gas drilling operations. He distinguished himself as an aggressive and highly competitive sportsman and was almost universally unpopular. He captured the gold medal in the 1948 Olympics in Britain sailing a six-meter yacht, and in 1977 he managed the *Independence-Courageous* yachting syndicate, which successfully defended the America's Cup. In 1946, Lee and Henry Loomis bought St. Vincent's Island off the coast of South Carolina, where they built a home for their mother. They eventually donated St. Vincent's to the Nature Conservancy. In 1963, Lee acquired Bull Island, where their family maintained a plantation and wildlife preserve until it was sold in

2000. He died in September 1994. He is survived by his four children: Nancy and Sabra, his wife's daughters by a previous marriage whom he adopted; Candace Stimson Loomis; and Alfred ("Chip") Lee Loomis III, an accomplished sailor and past commodore of the New York Yacht Club.

Farney Loomis became a respected biochemist and for many years was a professor at Brandeis University. With his father's backing, he opened his own Loomis Laboratories in Greenwich, Connecticut, where he conducted biological research on taming hydra. For several years, he and Alfred sponsored small scientific conferences to which they invited leading biologists, as in the old days at Tuxedo Park. When he was sixty years old, the depression that had plagued both his mother and his daughter caught up with him. He died of an overdose of sleeping pills in November 1973. He is survived by Joan and William Farnsworth Jr., his children by his first wife, Violet Amory; and an adopted daughter, Jacqueline, and Bart, his children with his second wife, Betty Loomis Evans. He later married Frances Whitman, adopted her three children, and had a son, Jefferson Loomis.

Henry Loomis went to work at the Berkeley Rad Lab immediately after the war, then shifted to MIT, where he worked as an assistant to the university's new president, James Killian, who was a friend of his father's. During World War II, when Stimson and Groves were trying to decide which Japanese cities to bomb, a chance visit by Henry, who had studied Japanese history at Harvard and had raved about the glorious art treasures in Kyoto, helped persuade Stimson to spare the ancient city. Henry pursued a career in public service on the advice of Stimson: "Uncle Harry used to tell us that we'd been kind of lucky in life and that we owed the country a duty." He served as chief of research and intelligence for the United States Information Agency and ran the Voice of America from 1958 to 1965. He later became president of the Corporation for Public Broadcasting. He is divorced from his first wife, Paulie Loomis. They had four children, Tim, Mary Paul ("Pixie"), Lucy, and Gordon. He is currently married to his second wife, Jacqueline, and lives in Jacksonville, Florida.

Henry Stimson's last cabinet meeting was on September 21, 1945, his seventy-eighth birthday. The day he left Washington to retire to his

Highhold estate in Long Island, he had lunch with Bush, and they drove out to National Airport together in his black limousine. "Every general officer in Washington was there, drawn up in lines from the car to the plane," recalled Bush. "General Marshall joined him and they walked down between those lines to the plane together. Never was a secretary more respected and revered." He received a nineteen-gun salute, and then the band played "Happy Birthday." His heart finally gave out in October 1950.

Landon K. Thorne continued in business with Loomis until 1955. After they finally liquidated the firm, each walked away with approximately $15 million. During World War II, Thorne was active in the Red Cross and several times was asked by Stimson, who was his wife's cousin, to carry important correspondence between the United States and Britain. Near the end of the war, during a dinner at their New York apartment, Stimson informed the party that the following day the United States was going to do something that would change the course of history. The next day, August 6, 1945, the first atomic bomb was dropped on Hiroshima. Landon Thorne died in 1964 and is buried in Woodlawn Cemetery in the Bronx, New York. Julia Atterbury Loomis Thorne died in 1973 and is buried beside her husband. She donated half of their 230-acre Bay Shore estate, Thorneham, to the Nature Conservancy. Landon Thorne Jr., who spent his career at Bankers Trust, died in 1980. Ed Thorne, a retired businessman, who lived in Greenwich, Connecticut, died in October 2002.

Honey Horn Plantation on Hilton Head Island, a total of twenty thousand prime acres of pine forest on the southern end, was sold by Loomis and Thorne in 1950 to a group of lumber associates from Hinesville, Georgia, called the Hilton Head Company. The property, which they had bought for $6 an acre, sold for $560 an acre—roughly $11.2 million, which Loomis later complained was too cheap. Two of the timbermen, General Joseph B. Fraser and Frederick C. Hack Sr., went on to develop the island as a vacation resort. The Hack family occupied Honey Horn until 1998, when the house and remaining sixty-eight-acre tract was purchased by the town of Hilton Head, which plans to restore the historic site and turn it into a museum.

Alfred L. Loomis' Scientific Publications

1927

Wood, R. W., and Alfred L. Loomis. "The Physical and Biological Effects of High-Frequency Sound-Waves of Great Intensity." *Philosophical Magazine* 4, no. 22 (September 1927).

Richards, William T., and Alfred L. Loomis. "The Chemical Effects of High Frequency Sound Waves." I. A Preliminary Survey. *Journal of the American Chemical Society* 49 (1927).

1928

Hubbard, J. C., and A. L. Loomis. "The Velocity of Sound in Liquids at High Frequencies by the Sonic Interferometer." *Philosophical Magazine* 5 (June 1928).

Loomis, A. L., and J. C. Hubbard. "A Sonic Interferometer for Measuring Compressional Velocities in Liquids: A Precision Method." *Journal of the Optical Society American Review of Scientific Instruments* 17 (October 1928).

Harvey, E. Newton, Ethel Browne Harvey, and Alfred L. Loomis. "Further Observations on the Effect on High Frequency Sound Waves on Living Matter." *Biological Bulletin* 55, no. 6 (December 1928).

1929

Richards, William T., and Alfred L. Loomis. "Dialectric Loss in Electrolyte Solutions in High Frequency Fields." *Proceedings of the National Academy of Sciences* 15, no. 7 (July 1929).

Harvey, E. Newton, and Alfred L. Loomis. "The Destruction of Luminous Bacteria by High Frequency Sound Waves." *The Journal of Bacteriology* 17, no. 5 (1929).

Christie, Ronald V., and Alfred L. Loomis. "The Relation of Frequency to the Physiological Effects of Ultra-High Frequency Currents." *Journal of Experimental Medicine* 49, no. 2 (February 1929).

Harvey, E. Newton, and Alfred L. Loomis. "A Chronograph for Recording Rhythmic Processes, Together with a Study of the Accuracy of a Turtle's Heart." *Science* 70, no. 1823 (December 1929).

1930

Loomis, Alfred L., E. Newton Harvey, and C. MacRae. "The Intrinsic Rhythm of the Turtle's Heart Studied with a New Type of Chronograph, Together with the Effects of Some Drugs and Hormones." *Journal of General Physiology* 14, no. 1 (September 1930).

Harvey, E. Newton, and Alfred L. Loomis. "A Microscope-Centrifuge." *Science* 72, no. 1854 (July 1930).

1931

Loomis, Alfred L. "The Precise Measurement of Time." *Monthly Notices of the Royal Astronomical Society* 140, no. 5 (1931).

Harvey, E. Newton, and Alfred L. Loomis. "High Speed Photomicrography of Living Cells Subjected to Supersonic Vibrations." *Journal of General Physiology* 15, no. 1 (September 1931).

1932

Loomis, Alfred L., and W. A. Marrison. "Modern Developments in Precision Clocks." *Bell Telephone Technical Publications* B656 (January 1932).

Loomis, Alfred L., and George Kistiakowsky. "A Large Grating Spectrograph." *Review of Scientific Instruments* 3 (January 1932).

Christie, Ronald V., and Alfred L. Loomis. "The Pressure of Aqueous Vapour in the Alveolar Air." *Journal of Physiology* 77, no. 1 (December 1932).

1933

Loomis, A. L., and H. T. Stetson. "An Apparent Lunar Effect in Time Determinations at Greenwich and Washington." *Monthly Notices of the Royal Astronomical Society* 93, no. 6 (June 1933).

1935

Loomis, A. L., and H. T. Stetson. "Further Investigations of an Apparent Lunar Effect in Time Determinations." *Monthly Notices of the Royal Astronomical Society* 95 (March 1935).

Loomis, Alfred L., E. Newton Harvey, and Garret Hobart. "Potential Rhythms of the Cerebral Cortex During Sleep." *Science* 81, no. 2111 (June 1935).

————. "Further Observations on the Potential Rhythms of the Cerebral Cortex During Sleep." *Science* 82, no. 2122 (August 1935).

1936

Loomis, Alfred L., E. Newton Harvey, and Garret Hobart. "Brain Potentials During Hypnosis." *Science* 83, no. 2149 (March 1936).

————. "Electrical Potentials of the Human Brain." *Journal of Experimental Psychology* 19, no. 3 (June 1936).

1937

Harvey, E. Newton, Alfred L. Loomis, and Garret A. Hobart III. "Cerebral Processes During Sleep as Studied by Human Brain Potentials." *Science* 85, no. 2210 (May 1937).

Davis, H., P. A. Davis, A. L. Loomis, E. N. Harvey, and G. Hobart. "Changes in Human Brain Potentials During the Onset of Sleep." *Science* 86, no. 2237 (November 1937).

Loomis, Alfred, E. Newton Harvey, and Garret A. Hobart III. "Cerebral States During Sleep, as Studied by Human Brain Potentials." *Journal of Experimental Psychology* 21, no. 2 (August 1937).

1938

Loomis, Alfred L., E. Newton Harvey, and Garret A. Hobart III. "Distribution of Disturbance-Patterns in the Human Electroencephalogram, with Special Reference to Sleep." *Journal of Neurophysiology* I (September 1938).

1939

Davis H., P. A. Davis, A. L. Loomis, E. N. Harvey, and G. Hobart. "A Search for Changes in Direct-Current Potentials of the Head During Sleep." *Journal of Neurophysiology* II (March 1939).

———. "Analysis of the Electrical Response of the Human Brain to Auditory Stimulation During Sleep." *American Journal of Physiology* 126, no. 3 (July 1939).

———. "Electrical Reactions of the Human Brain to Auditory Stimulation During Sleep." *Journal of Neurophysiology* II (November 1939).

PATENTS

Patent No. 1,376,890, issued May 3, 1921, to Alfred L. Loomis, Paul Klopsteg, Paul G. Agnew, and Winfield H. Stannard, for "Chronographs."

Patent No. 1,734,975, issued November 12, 1929, to Alfred L. Loomis and Robert Williams Wood, for "Methods and Apparatus for Forming Emulsions and the Like."

Patent No. 1,907,803, issued May 9, 1933, to E. Newton Harvey and Alfred L. Loomis, for "Microscope-Centrifuge."

Patent No. 2,884,628, issued April 28, 1959, to Alfred L. Loomis, for "Long Range Navigation System."

Author's Note
on Sources

ARCHIVES AND COLLECTIONS

The recollections of the early days of the Rad Lab, including those of Alfred Loomis, Luis Alvarez, Kenneth Bainbridge, Taffy Bowen, Karl Compton, Lee DuBridge, Donald Kerr, Frank Lewis, Ernest Pollard, and others quoted in the book, were taken from the interviews conducted by Henry Guerlac for the official history and belong to the Series: Records of the Office of the Historian; Subgroup: MIT Radiation Laboratory, RG 227, National Archives New England Region, Waltham, MA.

All letters written by Henry L. Stimson, as well as those written to him by Alfred L. Loomis, Ellen Farnsworth Loomis, and Julia Stimson Loomis, along with all the diary excerpts of Henry L. Stimson, are courtesy of the Henry L. Stimson Papers, Manuscripts and Archives, Yale University Library.

Grateful acknowledgment is made to the following:

Luis Alvarez: transcript of interview courtesy of the Center for History of Physics, the American Institute of Physics, New York.

Edward L. Bowles letters courtesy of the Library of Congress, Washington, D.C.

Vannevar Bush: oral history interviews courtesy of the Niels Bohr Library, the American Institute of Physics, College Park, Maryland. Letters courtesy of MIT and the Carnegie Institution of Washington.

Karl Compton: letters and papers courtesy of MIT.

James B. Conant: letters to Loomis courtesy of Harvard University Archives, Cambridge, Massachusetts.

Donald Cooksey: letter and papers courtesy of the Bancroft Library, the University of California, Berkeley.

Paul Klopsteg: transcript of interview courtesy of the Center for History of Physics, the American Institute of Physics, New York.

Ernest O. Lawrence: letters and papers courtesy of Ernest Orlando Lawrence Papers, the Bancroft Library, University of California, Berkeley. Also transcripts of taped interviews of Luis Alvarez and Alfred Loomis by Herbert Childs for his biography of Ernest O. Lawrence, *An American Genius*, published 1968.

John Lawrence: letters courtesy of the Bancroft Library, University of California, Berkeley.

Alfred Lee Loomis: Tuxedo Park laboratory guest book, letters, and papers courtesy of the Institute Archives and Special Collections, Massachusetts Institute of Technology Libraries, Cambridge, Massachusetts. Letters also courtesy of the Carnegie Institution of Washington.

William Richards: family letters and papers courtesy of Theodore R. Conant.

Leo Szilard: letters to and from Bill Richards, courtesy of Mandeville Special Collections Library, University of California, San Diego.

SELECT BIBLIOGRAPHY

Many scientists and historians have written in much greater detail about the events covered here. Among the publications to which I am indebted are the following:

Alvarez, Luis W. *Alvarez: Adventures of a Physicist*. New York: Basic Books, 1987.

———. "Alfred Lee Loomis." *Biographical Memoirs*, National Academy of Sciences, vol. 51, 1980.

———. "Alfred Lee Loomis—Last Great Amateur of Sciences." *Physics Today*, January 1983.

Amory, Cleveland. *The Last Resorts*. New York: Harper & Brothers, 1948.

———. *Who Killed Society*. New York: Harper & Brothers, 1960.

Anonymous. "Markets: Bonbright & Co., Inc." *Fortune*, February 1930.

———. "Niagara Hudson: Exhibit A of Superpower." *Fortune*, June 1931.

———. "The American Brain Barrel." *Fortune*, March 1945.

————. "Longhairs and Short Waves." *Fortune*, November 1945.

————. "The First Public Account of Radar." *Bureau of Naval Personnel Information Bulletin*, June 1943.

————. "Radar—The Industry." *Fortune*, October 1945.

————. "Alfred Lee Loomis: Amateur of the Sciences." *Fortune*, March 1946.

Baxter, James Phinney, III. *Scientists Against Time*. Boston: Little, Brown & Company, 1947.

Blackett, P. M. S. *Tizard and the Science of War*. Tizard Memorial Lecture, delivered to Institute for Strategic Studies, February 11, 1960.

Bowen, E. G. *Radar Days*. Bristol, Eng.: Hilger, 1987.

Brown, Louis. *A Radar History of World War II: Technical and Military Imperatives*. Bristol and Philadelphia: Institute of Physics Publishing, 1999.

Buderi, Robert. *The Invention That Changed the World*. New York: Touchstone, 1996.

Burns, Russell, ed. *Radar Development to 1945*. London: Peter Peregrinus/Institution of Electrical Engineers, 1988.

Bush, Vannevar. *Science—The Endless Frontier*. Washington, D.C.: U.S. Government Printing Office, 1945.

————. *Pieces of the Action*. New York: Morrow, 1970.

Chernow, Ron. *The House of Morgan*. New York: Touchstone, 1990.

Childs, Herbert. *An American Genius: The Life of Ernest Orlando Lawrence, Father of the Cyclotron*. New York: E. P. Dutton, 1968.

Churchill, Winston S. *The Gathering Storm*. Boston: Houghton Mifflin, 1948.

Clark, Ronald W. *Tizard*. Cambridge, Mass.: The MIT Press, 1965.

Clarke, Arthur C. *Glide Path*. New York: Harcourt Brace Jovanovitch, 1963.

Cockcroft, J. D. "Memories of Radar Research." *IEEE Proceedings* 132, pt. A, no. 6, October 1985.

Colton, Roger B. "Radar in the United States Army." *Proceedings of the IRE* 33, 1945.

Compton, Arthur Holly. *Atomic Quest: A Personal Narrative*. New York: Oxford University Press, 1956.

Conant, James B. *My Several Lives: Memoirs of a Social Inventor*. New York: Harper & Row, 1970.

Davis, Nuel Pharr. *Lawrence & Oppenheimer*. New York: Simon & Schuster, 1968.

Douglass, Paul F. *Six upon the World.* Boston: Little, Brown & Co., 1954.

Dowd, J., and George Lee. "A Scientist of Wall Street." *Popular Science Monthly,* August 1928.

DuBridge, L. A. "History and Activities of the Radiation Laboratory of the Massachusetts Institute of Technology." *The Review of Scientific Instruments* 17, no. 1 (January 1946).

DuBridge, L. A., and L. N. Ridenour. "Expanded Horizons." *Technology Review* 48, no. 1 (November 1945).

Goodchild, Peter. *J. Robert Oppenheimer: Shatterer of Worlds.* Boston: Houghton Mifflin, 1981.

Greenberg, Daniel S. *The Politics of Pure Science.* New York: New American Library, 1967.

Greer, Margaret. *The Sands of Time: A History of Hilton Head Island.* SouthArt, 1989.

Guerlac, Henry E. *Radar in World War II.* Los Angeles: Tomash Publishers; New York: American Institute of Physics, 1950.

Hartcup, Guy, and T. E. Allibone. *Cockcroft and the Atom.* Bristol, Eng.: Hilger, 1984.

Harvard College. "William Theodore Richards, 1900–1940." Decennial Report, Class of 1921. Cambridge, Mass., 1941.

Harvey, Natalie, ed. *Hilton Head Island: Images of America.* Charleston, S.C.: Arcadia Publishing, 1998.

Haskins, Caryl P. "Alfred Lee Loomis." Year Book of the American Philosophical Society, 1975.

Heilbron, J. L., and Robert W. Seidel. *Lawrence and His Laboratory: A History of the Lawrence Berkeley Laboratory,* Vol. 1. Berkeley: University of California Press, 1989.

Hershberg, James. *James B. Conant: Harvard to Hiroshima and the Making of the Nuclear Bomb.* New York: Knopf, 1993.

Hewlett, Richard G., and Oscar E. Anderson Jr. *The New World, 1939/1946: Vol. 1, A History of the United States Atomic Energy Commission.* University Park, Pa.: Pennsylvania State University Press, 1962.

Hodgson, Godfrey. *The Colonel: The Life and Wars of Henry Stimson, 1867–1950.* New York: Knopf, 1990.

Holmgren, Virginia C. *Hilton Head: A Sea Island Chronicle.* Hilton Head Island, S.C.: Hilton Head Island Publishing, 1959.

Holton, Gerald, ed. *The Twentieth-Century Sciences: Studies in the Biography of Ideas.* New York: W. W. Norton, 1972.

IEEE Center for the History of Electrical Engineering. *Rad Lab: Oral Histories Documenting World War II Activities at the MIT Radiation Laboratory*. Piscataway, N.J.: Institute of Electrical and Electronics Engineers, 1993.

Johnston, Lawrence. "The War Years." In Peter W. Trower, ed. *Discovering Alvarez: Selected Works of Luis W. Alvarez*. Chicago: University of Chicago Press, 1987.

Keegan, John. *The First World War*. New York: Knopf, 1999.

———. *The Second World War*. New York: Viking, 1989.

Kerr, Donald E., ed. *Propagation of Short Radio Waves*. Vol. 13, Radiation Laboratory Series. New York: McGraw-Hill, 1951.

Kevles, Daniel J. *The Physicists: The History of a Scientific Community in Modern America*. New York: Knopf, 1978.

Kistiakowsky, George B. *A Scientist at the White House*. Cambridge, Mass.: Harvard University Press, 1976.

———. Unfinished memoir. Courtesy of Elaine Kistiakowsky.

Lamont, Lansing. *Day of Trinity*. New York: Atheneum, 1965.

Lanouette, William, with Bela Silard. *Genius in the Shadows: A Biography of Leo Szilard, the Man Behind the Bomb*. New York: Charles Scribner's Sons, 1992.

Lescaze, William. "House at Tuxedo Park." *Architectural Review*, November 1939.

Loomis, Manette Seeldrayers. *From Belgium with Love*. Unpublished memoir courtesy of Al Hobart.

Marrison, Warren A. "The Evolution of the Quartz Crystal Clock." *The Bell System Technical Journal* 27 (1948).

Millar, D. A. "The Aberdeen Chronograph." *Army Ordnance* 4, no. 22 (January–February 1924).

Morison, Elting E. *Turmoil and Tradition: A Study of the Life and Times of Henry L. Stimson*. Boston: Houghton Mifflin, 1960.

Newton, Charles, Therma E. Patterson, and Nancy Joy Perkins. *Five Years at the Radiation Laboratory*. Cambridge, Mass.: Massachusetts Institute of Technology, 1946.

Office of Scientific Research and Development. "Radar: A Report on Science at War." 1948.

———. "Military Airborne Radar Systems: Vol. 2 of Summary Technical Report of Division 14." NDRC, 1946.

Pais, Abraham. *Niels Bohr's Times: In Physics, Philosophy and Polity*. Oxford, Eng.: Clarendon Press, 1991.

Powers, Thomas. *Heisenberg's War: The Secret History of the German Bomb*. New York: Knopf, 1993.

Randall, John T. "The Cavity Magnetron." Physical Society of London, *Proceedings* 58, pt. 3, 1946.

Rhodes, Richard. *The Making of the Atomic Bomb*. New York: Simon & Schuster, 1986.

Rich, Willard. *Brain Waves and Death*. New York: Charles Scribner's Sons, 1940.

———. "The Uranium Bomb." Unpublished manuscript courtesy of Ted Conant.

Rigden, John S. *Rabi: Scientist and Citizen*. New York: Basic Books, 1987.

Robinson, Forest G. *Love's Story Told: A Life of Henry A. Murray*. Cambridge, Mass.: Harvard University Press, 1992.

Seabrook, William. *Doctor Wood: Modern Wizard of the Laboratory: A Biography of Robert W. Wood*. New York: Harcourt, Brace & Co., 1941.

Segrè, Emilio. *A Mind Always in Motion: The Autobiography of Emilio Segrè*. Berkeley: University of California Press, 1993.

Skolnick, Merril I. "Fifty Years of Radar." *Proceedings of the IEEE* 73, no. 2 (February 1985).

Stimson, Henry L., and McGeorge Bundy. *On Active Service in War and Peace*. New York: Harper & Brothers, 1947.

Susskind, Charles. "Who Invented Radar?" In Russell Burns, ed., *Radar Development to 1945*. London: Peregrinus/Institution of Electrical Engineers, 1988.

Trower, Peter W., ed. *Discovering Alvarez: Selected Works of Luis W. Alvarez*. Chicago: University of Chicago Press, 1987.

Varian, Russell H., and Sigurd F. Varian. "A High Frequency Oscillator and Amplifier." *Journal of Applied Physics* 10 (May 1939).

Winslow, Albert Foster. *Tuxedo Park: A Journal of Recollection*. New York: Tuxedo Historical Society, 1992.

Winthrop, Stimson, Putnam & Roberts. *A History of the Law Firm*. New York, 1980.

Zachary, Pascal G. *Endless Frontier: Vannevar Bush, Engineer of the American Century*. New York: Free Press, 1997.

INTERVIEWS

Walter Alvarez

William J. Bozzuffi

George Boynton, Tuxedo Park
Lynn Chase, New York
Arthur Compton Jr.
Theodore R. Conant, New York
Martha H. Coolidge, Cambridge, Massachusetts
Peter Duchin
Betty Loomis Evans, East Hampton and New York
Janet Fisher, New York
Dr. John S. Foster, Connecticut
Charles Fraser, Hilton Head Island, South Carolina
Rowan Gaither III
Mimi Thorne Gilpatrick, New York
William Golden, New York
Patricia Gussin, Washington, D.C.
Frederick C. Hack Jr., Hilton Head Island, South Carolina
Natalie Harvey, Hilton Head Island, South Carolina
Caryl P. Haskins, Fairfield, Connecticut
Joan Loomis Hastings
Alfred Loomis Hobart, Waitsfield, Vermont
Garret Hobart IV, Waitsfield, Vermont
John Jessup, New York
Elaine Kistiakowsky, Cambridge, Massachusetts
Alfred Lee ("Chip") Loomis III, New York
Bart Loomis
William F. Loomis Jr.
Mary Paul "Paulie" Loomis, Massachusetts
Tim Loomis, who interviewed his ailing father, Henry Loomis
Sabra Loomis, New York
Talbot Love, Tuxedo Park Historical Society
John Modder, Tuxedo Park
John Jay Mortimer, New York
Philip Nash, New York
Mary Lawrence Prudhomme
Jacqueline Loomis Quillen, East Hampton and New York
Christian Sonne, Tuxedo Park
Edwin Thorne, Greenwich, Connecticut
Landon Thorne III, Beaufort, South Carolina
Richard H. Tourin

Acknowledgments

This book could not have been written without the encouragement of my parents, especially my father, who first planted the seeds of this story years ago and, when I asked, willingly turned over his family's papers and letters and spent hours rummaging through boxes and stacks of books. I am also indebted to Betty Loomis Evans and her children, Jacqueline Loomis Quillen and Bart Loomis, who welcomed a stranger with open arms and gave so generously of their time. They shared their detailed memories of their family's past with great honesty and insight, provided old photographs, and steered me to longtime friends and colleagues. I want to thank, in particular, Alfred and Garret Hobart, for their reminiscences and for giving me access to their mother's records and photographs. With few exceptions, all the members of the extended Loomis/Thorne clan were forthcoming, and without their help this book would not have been possible.

Most of Alfred Loomis' papers were lost or discarded when he sold the Tower House in Tuxedo Park in 1950. The few remaining laboratory records, along with the historic guest book, are deposited in the Institute Archives and Special Collections room at MIT. Fortunately, Henry Stimson was an avid letter writer and filled volumes of diaries with his thoughts and observations, and a rich store of material is available in the archives of Yale University Library. Digging up material on Loomis' myriad activities involved a considerable amount of detective work, and I owe an immense debt to my researcher, Ruth Tenenbaum. She is persistent, patient, and resourceful and came through no matter how obscure the request. I could not ask for a better colleague and friend.

I wish to thank Elaine Kistiakowsky for her kind assistance, for providing a bunk in Cambridge, and for entrusting me with George's unpublished memoir. Spending time with her again was like coming home in more ways than one. The late Caryl Haskins guided me through Alfred Loomis' early scientific career and served as an invaluable sounding board. William Golden graciously took the time to review the manuscript, and I benefited from his perceptive comments and wise counsel. For their time and informative tours of Tuxedo Park, thanks also to George Boynton and Chris Sonne. Donna Moreau did yeoman's work going through the massive Stimson archive.

For moral and editorial support, I owe an enormous debt of gratitude to my friends Barbara Kantrowitz and Daniel Hertzberg for their help on the work in progress, for cheering me on, and for their much appreciated advice at every stage.

I count myself fortunate to have the most caring and exacting editor in Alice Mayhew at Simon & Schuster. We talked about my writing a book for many years, and without her encouragement and infectious enthusiasm I never would have settled down to the task. At S&S, I also benefited from the careful pencil of Roger Labrie, the rigorous copy editing of Sona Vogel, and the attention of Alice's assistants, Anja Schmidt and Jonathan Jao. My deepest personal thanks are owed to my agent, Kris Dahl, at ICM, who has been a great friend and adviser for more than a decade.

And most of all I want to thank my husband, Steve Kroft, for his love and understanding over the long haul. I could not have done it without him. I also want to thank my son, John, for always finding his way clear of all the papers on the floor to give me a kiss when he got home. It made the solitary days bearable.

Index

Photo Credits